Investigating Probability and Statistics Using the TI-81 and TI-82 Graphics Calculators

Teacher's Guide

Graham A. Jones
Roger Day
Beverly S. Rich
Carol A. Thornton

Innovative Learning Publications

Addison-Wesley Publishing Company

Menlo Park, California • Reading, Massachusetts • New York
Don Mills, Ontario • Wokingham, England • Amsterdam • Bonn
Paris • Milan • Madrid • Sydney • Singapore • Tokyo
Seoul • Taipei • Mexico City • San Juan

We wish to thank Michelle Bell, Elaine Caspers, Dana Jensen, Marion Jones, Sheryl Kidd, and Marilyn Parmantie, all of whom contributed to the books in a variety of ways.

Managing Editor: Cathy Anderson
Project Editor: Katarina Stenstedt
Production: Leanne Collins
Design Manager: Jeff Kelly
Text and cover design: London Road

This book is published by Innovative Learning Publications™, an imprint of the Alternative Publishing Group of Addison-Wesley Publishing Company.

ISBN 0-201-49406-X

3 4 5 6 7 8 9 10-ML-99

Contents

To the Teacher

In the *Curriculum and Evaluation Standards for School Mathematics* (1989), the National Council of Teachers of Mathematics focuses national attention on new goals for school mathematics. This document includes identification of specific standards for probability and statistics at all levels. In accord with these recommendations, *Investigating Probability and Statistics Using the TI-81 and TI-82 Graphics Calculators* adopts a contemporary approach to data and chance that incorporates the use of technology.

Statistics emphasizes "making sense of data" through exploring, organizing, and interpreting relevant data in a variety of ways. Ideas developed in statistics are linked to probability. Probability emphasizes simulating real-world problems that involve students in experimenting, collecting, organizing, and using data.

Investigating Probability and Statistics Using the TI-81 and TI-82 Graphics Calculators is intended to be used as a supplement to the regular instructional program, and most topics will need some preliminary introduction by you, the teacher. With this in mind an extensive section on content background with many worked examples has been provided for each of the four modules.

SPECIAL FEATURES OF THIS TEACHER'S GUIDE

The Teacher's Guide for each of the four modules of Investigating Probability and Statistics Using the TI Graphics Calculators incorporates the following helpful features:

- **Overview and Purpose.** This identifies the spirit and scope of the module and the expectations to be set for students.

- **Outline of the Key Mathematical Ideas by Activity.** This is provided as a ready reference during planning and implementation of the instructional program. The Calculator Help procedures germane to each module are also listed.

- **Content Background.** These detailed sections use a problem-solving approach in order to accommodate the diverse backgrounds of teachers. The material is also presented in a way so that much of it can be used as part of the instructional program. This section includes reference to other resources that provide useful background on the module topics. A full listing of these resources is included at the end of this Teacher's Guide.

- **Classroom Activities.** These involve tasks, problems, or games that capture students' interests and elicit questions that are relevant to the content and expectations of the modules.

- **Implementation Guidelines.** Activity by activity, these sections provide specific suggestions for planning and implementing the instructional program and assessing student performance.

In addition to these features selected answers are provided for key problems.

FEATURES OF THE STUDENT EDITION

There are four modules in the Student Edition: Analyzing Data; Probability, Simulations, and Random Numbers; Counting; and Modeling and Predicting. The *Looking Ahead* section at the beginning of each module overviews the kinds of activities students will explore during the module. A *Looking Back* section concludes each module to help students assess their understandings of the key concepts in the module and share their thoughts with others.

Each module has three levels of activity—Starting, Developing, and Extending—to accommodate diverse students. Each level includes a series of activities and appropriate calculator instructions. If your students have had little or no experiences in probability and statistics, begin with the Starting activities. If they have had substantial experience they may be able to move directly to the Developing activities or even to the Extending activities. In any case, wherever students start they may need to check back on some of the technology techniques introduced at various stages of the book.

To reflect new approaches to the teaching of data and chance, each activity contains three major sections: *Organize, Communicate,* and *Reflect.* A focus problem is examined as students experiment, *organize,* and analyze data. The outcomes of their analyses are *communicated* through written and verbal interactions. Finally, the *Reflect* section poses questions that encourage students to summarize key ideas and to extend their understandings.

The Student Edition contains a number of special features including calculator help pages, calculator programs, data sets, a word bank, blackline masters, and required materials boxes. More details on these features are provided in the *Implementation Guidelines* section in each module of the Teacher's Guide and in the Student Edition.

MODULE 1:
Analyzing Data

OVERVIEW AND PURPOSE

The activities in this module emphasize organizing, describing, summarizing, and analyzing data. As students explore these activities with a TI graphics calculator, they should achieve greater insights into the patterns, trends, and anomalies of data. In essence the goal of this module is to develop "data sense." Students will learn how to:

- organize data in meaningful ways;
- describe and summarize data using measures of location, spread, and shape;
- use technology to construct visual presentations of data; and
- analyze and interpret data.

While the activities enhance and apply key concepts used in TI graphics calculators for describing and analyzing data, they are not intended to provide the initial development of these concepts. Because you know the backgrounds and learning experiences that students bring to their study of data analysis, you are best suited for this role.

With this in mind, the following sections provide:

- an outline of key mathematical ideas by activity;
- content background;
- sample classroom activities; and
- guidelines for implementing the activities of Module 1.

OUTLINE OF KEY MATHEMATICAL IDEAS BY ACTIVITY

Activity Title	Key Mathematical Ideas
Starting	
1-1　What's in a Name?	collecting, entering data
1-2　Names in a Histogram	histogram
1-3　What's the Mean for the "Name" Data?	mean
1-4　What's the Median for the "Name" Data?	median
1-5　Quartering the "Name" Data	quartiles
1-6　What's the Mode for the "Name" Data?	mode
1-7　Measuring the Spread of the "Name" Data	range, interquartile range
1-8　Who's Far Out?	outliers

(Outline of Key Mathematical Ideas, continued)

Developing

1-1	Games Won	stem-and-leaf plot
1-2	Runs Scored	stem-and-leaf plot
1-3	Truncate or Round?	truncating, rounding
1-4	Comparing Runs	back-to-back plots
1-5	Boxing Games Won	box-and-whisker plot
1-6	Comparing Games Won	multiple box plots
1-7	Runs Scored and Games Won	scatter plot
1-8	Names and Shoe Lengths	correlation

Extending

1-1	Roll a Dozen	analyzing the die data
1-2	Who Was the Top Female Tennis Player?	interpreting data
1-3	The Top Ten Roadsters	describing data
1-4	The Speeds of Animals	analyzing data
1-5	Endangered Animals	analyzing data

Calculator Helps

For Module 1:
Calculator Help 1: Entering Data
Calculator Help 2: Range; Histogram
Calculator Help 3: Mean
Calculator Help 4: Generating a Scatter Plot

CONTENT BACKGROUND

One-Variable Data

Looking at data in their raw form, it is difficult to detect patterns. An unorganized data set does not lend itself to appropriate description. For example, the 1994 *Information Please Almanac* provides data on "Motion Picture Revenues: All Time Top Money Makers." Thirteen of these, together with the total rental revenue for each film, in millions of dollars, are presented below.

Figure 1
Motion Picture Revenues: All Time Top Money Makers *(amounts are in millions of dollars)*

Batman	$150
Empire Strikes Back	142
E.T. The Extra-Terrestrial	228
Ghostbusters	133
Grease	96
Home Alone	140
Indiana Jones and the Last Crusade	116
Jaws	130
Raiders of the Lost Ark	116
Rainman	87
Return of the Jedi	169
Star Wars	194
Terminator 2	112

Source: Information Please Almanac, 1994

One way to organize the data set is to list it in descending order. Another way is to group the data into appropriate clusters. Both of these are incorporated into the list in Figure 2.

Figure 2

Motion Picture Revenues: All Time Top Money Makers *(amounts are in millions of dollars)*

E.T. The Extra-Terrestrial	$228	Cluster I
Star Wars	194	Cluster II
Return of the Jedi	169	
Batman	150	
Empire Strikes Back	142	
Home Alone	140	
Ghostbusters	133	
Jaws	130	
Raiders of the Lost Ark	116	
Indiana Jones and the Last Crusade	116	
Terminator 2	112	
Grease	96	Cluster III
Rainman	87	

Ordering a data set in this way is a useful first step in summarizing the data set and generating visual representations of it. In a one-variable data set, such as the Motion Picture Revenues data presented here, summaries of the data can be conveyed through the essential characteristics of *location, spread,* and *shape.* These are the key considerations in the Starting activities of Module 1.

CENTRAL LOCATIONS

In describing the location of the data, determine one or more representative values that either locate the center of the data (the median and the mean do this) or identify a typical value in the data (the mode does this).

Figure 3 provides examples of how to determine these three key measures of location. Provided with the examples are comments and references that amplify some characteristics of these key measures.

SPREAD

To describe the spread of a data set is to characterize the variability or dispersion of the values in the data set. Measures of spread typically are based on position. The *range* describes the spread based on the minimum and maximum values of the data. On the other hand the *interquartile range* characterizes the spread of the data based on the upper and lower quartiles of the data set. Figure 4 demonstrates how to calculate these two measures of spread.

Figure 3

Central Locations: Motion
Picture Revenues Data

Median: middle value

87, 96, 112, 116, 116, 130, $\boxed{133}$ 140, 142, 150, 169, 194, 228

$m = \$133,000,000$

Mode: most frequently occurring value

mode = $116,000,000

Mean: arithmetic average

$$\text{mean} = \frac{(87+96+112+116+116+130+133+140+142+150+169+194+228)}{13}$$

$= 139.5$

$= \$139,500,000$

Notes:

1. *The median is also called the 50th percentile.*

2. *If the first value, $87 million, was omitted, there would be an even number of scores. The median would then be the mean of $133 million and $140 million, that is, $136.5 million.*

3. *For some data sets, there may be more than one mode. In these cases, the data are multimodal.*

4. *The mean is affected by extreme scores. To see the impact, replace $228 million with $428 million and recalculate the mean, median, and mode.*

5. *References: [4], [5], [6], [8], [11], [12]*

Figure 4

Measures of Spread:
Motion Picture
Revenues Data

Range: the difference between the maximum and minimum data values

= 228 (upper extreme) − 112 (lower extreme)

= $116,000,000

Interquartile Range (IQR): the difference between the 75th and 25th percentiles

Step 1: Find the median (the 50th percentile) of the data set and then find the median of each half of the data set. The median of the lower half is the lower quartile (the 25th percentile) and the median of the upper half is the upper quartile (the 75th percentile).

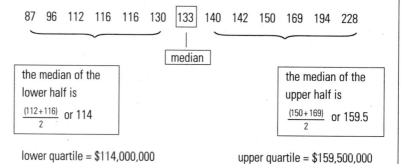

lower quartile = $114,000,000   upper quartile = $159,500,000

Step 2: Subtract the lower quartile from the upper quartile.

159.5 − 114

interquartile range = $45,500,000

Notes:

1. *When determining the lower quartile as the median of the lower half of the data and the upper quartile as the median of the upper half of the data, first omit the median of the entire data set. In the example above, 133, the median, is omitted.*

2. *Two other measures of spread, the variance and the standard deviation, are not considered in this guide. See references [5] and [11] for specifics of these and other measures of spread.*

3. *References: [4], [5]*

SHAPE

The reference for characterizing shape is the normal or mound-shaped distribution. This is a perfectly symmetrical distribution where median, mode, and mean are the same value.

Shapes that deviate from a symmetric distribution often have extreme values or outliers. Extreme values affect the mean of the data but not the median or the mode. (See Figure 5.)

Figure 5

Shapes of Distributions

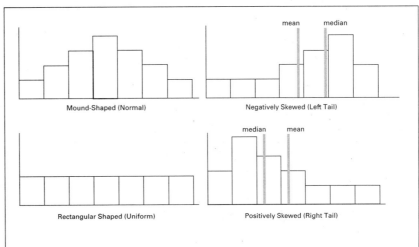

Notes:

1. A mound-shaped or normal distribution is symmetrical and the mean, the median, and the mode have the same value. Height in the population at large produces a mound-shaped distribution.

2. Skewed distributions tail to the left (negative) or to the right (positive). Extreme values, including outliers at one end of the distribution, affect the mean but not the median. The extent of difference between the mean and the median determines the direction and magnitude of skewness in a distribution. Heights of sample populations that contain some professional basketball players would produce a positively skewed distribution.

3. In a rectangular distribution data values are uniformly spread across the values of the variable. The outcomes of a single roll of a die, repeated many times, produce a rectangular distribution.

4. References: [4], [5]

PRESENTATIONS OF DATA

In Module 1 students construct and interpret histograms, stem-and-leaf plots, and box-and-whisker plots. Each of these visual displays has distinctive features that make it valuable for presenting different kinds of data.

The key feature of a histogram is that it displays the frequency, or count, of each value or class of values in the data set. Figure 6 shows a histogram for the Motion Picture Revenues data.

Figure 6

Histogram: Motion Picture Revenues Data

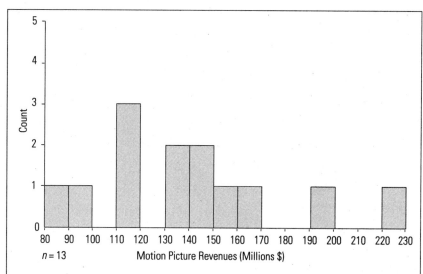

Notes:

1. The data have been truncated. Thus, the values 112, 116, and 116 are represented by 110.

2. The frequency, or count, is shown for each truncated data value, or data class. Often histograms use relative frequencies. For example, the relative frequency for the $110 million truncated data would be $\frac{3}{13} = 0.23$ or 23 percent.

3. In this guide students use single values in each class. With more complex data, a data class may contain a range of data values, such as greater than or equal to 80 million and less than 100 million.

A stem-and-leaf plot uses the decimal system's place values to generate classes of data values. Stem-and-leaf plots have the added advantage of preserving all the values of a data set.

A box-and-whisker plot is based on a five-number summary that situates position markers among the data values. The five-number summary includes the maximum and minimum data values (the *upper extreme* and the *lower extreme*), the 25th and 75th percentiles (the *lower quartile* and the *upper quartile*), and the 50th percentile (the *median*). Figure 7 presents stem-and-leaf plots and box-and-whisker plots for the Motion Picture Revenues data.

Figure 7

Stem-and-Leaf Plots and
Box-and-Whisker Plots:
Motion Picture Revenues
Data

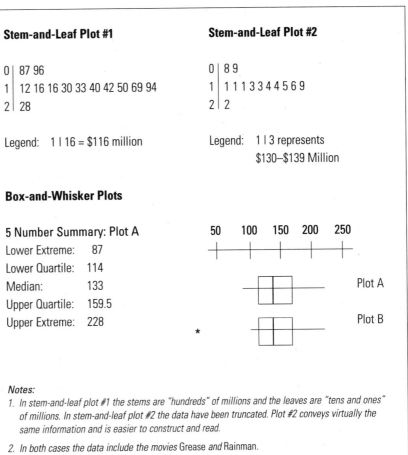

Stem-and-Leaf Plot #1

```
0 | 87 96
1 | 12 16 16 30 33 40 42 50 69 94
2 | 28
```

Legend: 1 | 16 = $116 million

Stem-and-Leaf Plot #2

```
0 | 8 9
1 | 1 1 3 3 4 4 5 6 9
2 | 2
```

Legend: 1 | 3 represents
          $130–$139 Million

Box-and-Whisker Plots

5 Number Summary: Plot A
Lower Extreme: 87
Lower Quartile: 114
Median: 133
Upper Quartile: 159.5
Upper Extreme: 228

Notes:

1. *In stem-and-leaf plot #1 the stems are "hundreds" of millions and the leaves are "tens and ones" of millions. In stem-and-leaf plot #2 the data have been truncated. Plot #2 conveys virtually the same information and is easier to construct and read.*

2. *In both cases the data include the movies* Grease *and* Rainman.

3. *Plot A represents the original data set. Plot B is the data set if* Boomerang *(revenue: $34,000,000 replaced* Rainman *(revenue: $87,000,000). The data have an outlier. In a box-and-whisker plot, outliers are those values more than 1.5 times the IQR (Interquartile Range) beyond either the upper or the lower quartile. Such an outlier is represented with an asterisk as shown in box-and-whisker-plot #2.*

4. *References: [4], [5], [8], [11], [12]*

Two-Variable Data

For a two-variable, or *bivariate,* data set, the essential characteristics are *direction, strength,* and *shape.* Concepts and skills associated with these characteristics can help effectively describe a data set. Activities in Module 1 help students create *scatter plots* of bivariate data. A scatter plot provides a first look at how two sets of data may relate. It provides a visual display of the data that may reveal characteristics of direction, strength, and shape not apparent from the raw data. Scatter plots will be especially helpful in Module 4 when students find the lines and curves of best fit to model two-variable data.

DIRECTION

Scatter plot (a) of Figure 8 shows the relationship between motion picture revenues (*x*) and ranking (*y*). Observe the direction of this relationship. Notice that as rental revenues increase, the ranking improves, which in this case means the ranking numbers decrease. The direction of this relationship is said to be *negative*.

If you had considered the relationship between motion picture revenues (*x*) and the number of people renting the movies (*y*), the direction of this relationship would be positive, because as revenue increases, so does the number of people renting the movie. A scatter plot of this relationship might appear as plot (b) in Figure 8.

On the other hand, if you consider the relationship between motion picture revenues (*x*) and the shoe size of each movie's executive producer (*y*), it is likely that the relationship will be neither positive nor negative. In this case there is no apparent direction to the relationship, as is illustrated in plot (c) of Figure 8.

Figure 8

Relationships Revealed in Scatter Plots

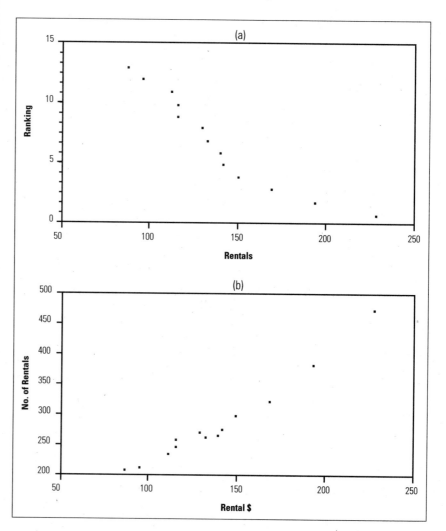

Figure 8 (cont.)

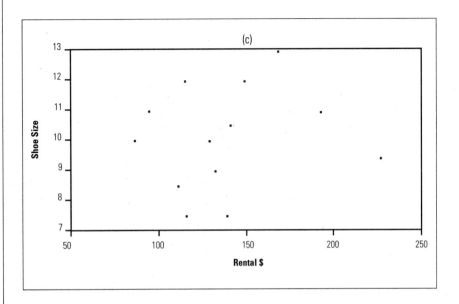

STRENGTH

To look at the strength of the relationship in a two-variable set of data, it is customary to apply the ellipse test. In Figure 9 the ellipse test is applied to each of the scatter plots examined in Figure 8. In each case an ellipse fully captures the plotted points. To assess the strength of a relationship, consider the shape of the ellipse. In plots (a) and (b) each ellipse is long and narrow, indicating a moderate to strong linear relationship in each case. In plot (c) the ellipse may be described as a fat ellipse or even a circle. This indicates a very weak or nonexistent linear relationship between the two variables.

In Module 4 students will also measure the strength of linear relationships using a *correlation coefficient.* The correlation coefficient is related to both the shape of the ellipse and the direction of the relationship. For example, plot (a), with its narrow ellipse and negative direction, has a correlation coefficient close to −1. Plot (b), which exhibits the same strength of linear relationship but has positive direction, has a correlation coefficient close to +1. Appropriately, plot (c), with its fat ellipse and no apparent direction, has a correlation coefficient close to 0.

SHAPE

Scatter plots that show a linear relationship illustrate only one of the shapes that model two-variable data. Figure 10 shows scatter plots of two-variable data sets that are quadratic (a), exponential (b), and logarithmic (c). Students will explore all these shapes in Module 4.

Figure 9

Ellipse Test Applied to
Scatter Plots

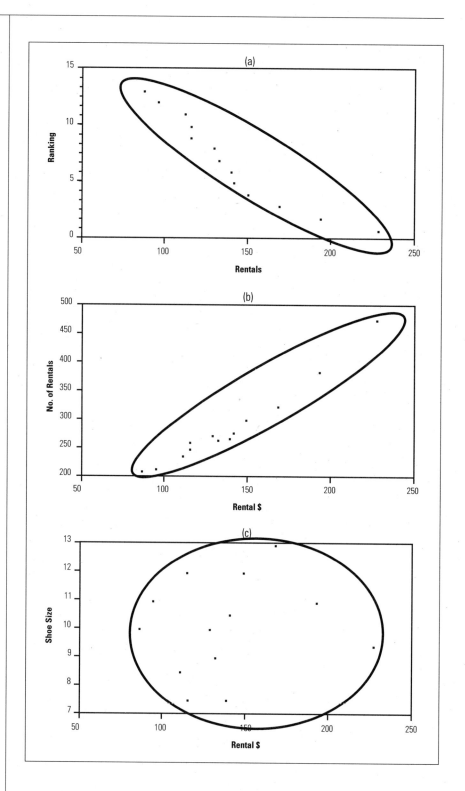

Figure 10

Scatter Plots of
Two-Variable Data Sets

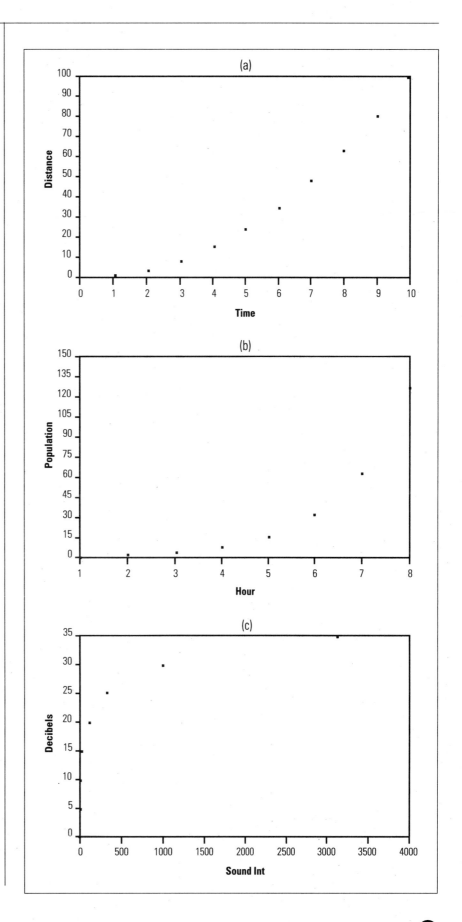

**CLASSROOM
ACTIVITIES**

Use the following activities to explore a number of the statistical concepts examined in Module 1.

Activity 1—Median and Quartiles

Measure student heights in centimeters. Arrange students in a line so they are ordered from tallest to shortest. Determine values for the median, the lower quartile, and the upper quartile.

This activity also can be done by having each student write his or her height on an index card. Ask students to order the cards from the largest value to the smallest value.

Activity 2—Stem-and-Leaf Plot

Use the data generated in Activity 1. Have students create a life-size stem-and-leaf plot using the hundreds and tens digits as the stems and the ones digits as the leaves. For example, a height of 164 cm would be represented as 16 | 4. Create the stems with appropriate index cards and have each student become a leaf by holding a card showing the ones digit corresponding to his or her height.

Have students locate the positions of the median and quartiles on the life-size plot. Note that this life-size plot also represents a histogram in which the data class width is 10 cm. A typical data class is 160 cm to 169 cm. (Self-adhesive notes on the chalkboard also work well.)

Activity 3—Box Plot 1

Again using the data generated in Activity 1 and the same physical lineup of students, find the elements of the five-number summary, the range, and the interquartile range. Use tape on the classroom floor to construct a box-and-whisker plot based on the student lineup. Examine the plot and determine whether outliers exist.

Activity 4—Box Plot 2

Measure student shoe lengths in centimeters. Arrange the male students and the female students in separate lines each ordered from the shortest shoe length to the longest. Find the elements of the five-number summary for each group. Use tape on the classroom floor to construct box-and-whisker plots for each group. Compare locations, spread, and shape.

Activity 5—Arithmetic Mean

The purpose of this activity is to demonstrate that the arithmetic mean has interpretation beyond the formula $\frac{\text{sum of the data values}}{\text{number of data values}}$.

As one interpretation, the mean should be recognized as the unique number that could replace each value in the data set and still result in the data having the same sum. The mean can also be interpreted as the balancing point of the data. This can be visualized with a number line, as below. Consider the data set 9, 7, 6, and 2. Its mean is 6. Note that the value 9 is three units to the right of the mean and the value 7 is one unit to the right of the mean. Together, these two data values are 4 units to the right of the mean. The value 6 is at the mean. Moreover, the value 2 is four units to the left of the mean and consequently balances the effect of the other data values. This balance-point characteristic of the mean also demonstrates that the average deviation from the mean is 0 for any data set.

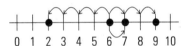

To emphasize the concept of the mean, select a number that is a divisor of the number of people in the class. (If your class has a prime number of students, you should join the class for this activity.) Have students form that many groups so that not all groups have the same number of students. For example, if there are 24 students in the class, you could form 2, 3, 4, 6, 8, or 12 groups, with varying numbers in each group. For instance, if 4 groups are formed, you might have groups of 9, 7, 6, and 2 students.

Ask students to interchange group members so that each group has the same number of students. In the example given here, this number is 6, the mean number of students in the original groups. Ask the students to discuss the properties of the mean exhibited by the activity, including how the mean relates to the total number of students, to the number of groups formed, and to the original size of each group.

IMPLEMENTATION GUIDELINES

This section contains suggestions for supporting an instructional program based on the activities contained in the Student Edition. For most activities comments are provided that might be helpful in preparing, implementing, or assessing the activity.

Resources

A *Need* list is located at the bottom of each activity page. Check this carefully because it indicates what information or resources are needed for the activity. It refers to data sets, blackline masters, and calculator programs, which are found in the appendices of the Student Edition, as well as to dice, coins, or other manipulatives, which are supplied by you or students.

On the activity page students are also referred to the page in Appendix B of the Student Edition where they can find relevant calculator help information. The Student Edition also contains a word bank (Appendix E) of common terms used in probability and statistics.

Convention for Using the 2nd or Alpha Keys

The convention adopted throughout this guide is that, whenever the 2nd or ALPHA keys are pressed, the colored label above the next key to be pressed is stated. For example, when using [STAT] (above the key MATRX) with the TI-81, the instruction will be 2nd [STAT].

Teaching Suggestions by Activity

Starting 1-1 What's in a Name? This activity is typical of several activities in this guide that allow students either to collect their own data or to use data sets that are provided. Activities of this kind are consistent with various reform recommendations that suggest students need repeated experiences in gathering and organizing data.[1] This first activity focuses on sorting and ordering data. This is one of the preliminary means of organizing data for developing visual presentations such as histograms.

---

[1]National Council of Teachers of Mathematics (NCTM), *Curriculum and Evaluation Standards for School Mathematics,* 1989; National Research Council, *Everybody Counts: A Report to the Nation on the Future of Mathematics Education,* National Academy Press, 1989.

Starting 1-2 Names in a Histogram This activity introduces ideas about clusters, gaps, and outliers. Only an intuitive discussion is suggested at this stage, as these topics will be considered in more detail later. Do this activity as close as possible to Starting 1-1 so students do not have to reenter data. The *Need* list indicates that the data from Starting 1-1 are needed. Watch for this in other activities throughout the Student Edition.

Starting 1-3 What's the Mean for the "Name" Data? Some of the classroom activities presented earlier would provide a useful introduction to this activity. When determining the mean, it may be helpful for students to work in pairs—one using the calculator while the other reads the values.

Starting 1-4 What's the Median for the "Name" Data? If students are not familiar with subscripts, it will be necessary to illustrate and discuss them, for these appear on the calculator screen and will be used to find the median.

Starting 1-5 Quartering the "Name" Data Students find the lower quartile as part of this activity. For assessment, ask students to find the upper quartile.

Starting 1-6 What's the Mode for the "Name" Data? Before beginning this activity, students who are unfamiliar with "frequency" should be referred to Appendix E, "Word Bank," in the Student Edition.

Starting 1-7 Measuring the Spread of the "Name" Data This activity considers only two measures of spread—the range and the interquartile range. If you wish to introduce variance and standard deviation, these can be found on the calculator using <1-Var> under the STAT CALC menu.

Developing 1-1 Games Won This activity focuses on stem-and-leaf plots, one of several ways to present data visually. The leaves do not have to be ordered, but it is useful to do so. The stem-and-leaf plot is particularly helpful for finding the median, mode, and quartiles. This is the first activity that uses a program for the TI-82 calculator (see Appendix C, "Calculator Programs," in the TI-82 Student Edition). Be sure to read the introduction to this appendix.

Developing 1-2 Runs Scored This activity displays three-digit numbers in a stem-and-leaf plot. Some discussion of the way to handle more than two digits (typically by placing only the last digit in the leaf) is appropriate.

Developing 1-3 Truncate or Round? Truncating involves dropping digits, whereas rounding follows the usual rules. An illustration and discussion of the distinction between these two techniques will be helpful before using the ideas in constructing stem-and-leaf plots. For further details see reference [4].

Developing 1-4 Comparing Runs This activity introduces back-to-back stem-and-leaf plots. The calculator can be used to order the data prior to constructing the stem-and-leaf plot by using the "sort" feature of the TI graphics calculator, described in the calculator manual. This type of plot allows the student to see clusters as well as bimodal or skewed characteristics of the data.

Developing 1-5 Boxing Games Won This activity introduces box-and-whisker plots. Because students will need to determine the median, quartiles, and extremes in order to construct the plots, a review of these terms may be helpful (see Appendix E, "Word Bank," in the Student Edition). After constructing the graph, students will see how the term "box-and-whisker" was derived.

Developing 1-6 Comparing Games Won In a box-and-whisker plot the median does not always fall in the middle of the box. The median is the value below which 50 percent of the data values fall and above which 50 percent of the data values fall. The endpoints of the box are formed by the upper and lower quartile values, which do not have to be equidistant from the median. A discussion of these properties would be helpful, especially using skewed data.

Developing 1-7 Runs Scored and Games Won This activity introduces a scatter plot. Instruct students to clear the data before entering new values. (See "Calculator Help 1: Entering Data" in Appendix B of the Student Edition.) Students also should check the range before graphing (see "Calculator Help 2: Range; Histogram" in Appendix B of the Student Edition) to be sure the range of the data "fits" the assigned range values. A discussion of independent and dependent variables (see Appendix E "Word Bank" in the Student Edition) is appropriate.

Developing 1-8 Names and Shoe Lengths Discuss the notion of correlation (see Appendix E "Word Bank" in the Student Edition) and, in particular, linear correlation (see *Content Background* on page 9).

Extending 1-1 Roll a Dozen In Extending activities such as this, you should expect students to work more independently. You may want to use these activities as assessment projects.

Extending 1-2 Who Was the Top Female Tennis Player?
Students should decide whether to truncate or round, though it might be appropriate to raise the question about which they prefer to use and why. Student graphs might make an attractive wall display. Some students might even like to gather data on men's tennis as an extension of this activity.

Extending 1-3 The Top Ten Roadsters This activity requires students to analyze the data set in order to determine how best to enter it into the calculator. If they use 1991 and 1990 for the scatter plot, the same cars are listed (see "Data Set 5: The Ten Top-Selling Passenger Cars in the United States by Calendar Year (1989–1991)" in Appendix D of the Student Edition). However, if the 1989 cars are added a reduced set of 1990 and 1991 cars is necessary.

Extending 1-4 The Speeds of Animals and Extending 1-5 Endangered Animals Challenge students to consider various visual presentations and to evaluate which visual display is most effective in interpreting the data.

MODULE 2:
Probability, Simulations, and Random Numbers

OVERVIEW AND PURPOSE

The activities in Module 2 focus on real-world problems of chance. The problems can be effectively modeled with spinners, dice, coins, random numbers, or other probability generators. By simulating real-world activities with probability devices, students determine the *experimental probabilities* involved. The TI graphics calculators not only allow students to generate random numbers in various groupings, but also to graph experimental results and to determine the mean number of times an event occurs in a series of trials. Students will learn how to:

- use simple counting techniques and measures such as area to determine theoretical probabilities;

- collect and use appropriate data to determine experimental probabilities;

- use experimental probabilities to predict theoretical probabilities; and

- model and simulate problems involving chance using probability devices, especially random numbers.

The calculator-supported activities in this section are designed to provide rich development in experimental probability and in the use of simulation for solving problems involving chance. However, you will need to help students explore a number of basic concepts, especially those associated with theoretical probability, prior to undertaking the activities in this module. The following sections are intended to provide background for this preliminary study and provide greater insights into the simulation process used in problem solving.

With this in mind, the following sections provide:

- an outline of key mathematical ideas by activity;

- content background;

- sample classroom activities; and

- guidelines for implementing the activities of Module 2.

OUTLINE OF KEY MATHEMATICAL IDEAS BY ACTIVITY

Activity Title	Key Mathematical Ideas
Starting	
2-1 Flip the Coin	experimental probability
2-2 Model the Coin Flip	using random numbers
2-3 Modeling Two Flips	two-stage experiments
2-4 Steffi's Serve	modeling
2-5 Flip the Cup	experimental probability
2-6 Chances for the Name	data, experimental probability
2-7 Modeling the Chances for the Name	modeling
2-8 Two Broadway Favorites	modeling a two-stage experiment
2-9 Three in a Row	three-stage experiments
Developing	
2-1 Three Hits	modeling
2-2 The Top Ten	modeling
2-3 On Top for Four Weeks	modeling
2-4 Basketball Playoffs	modeling
2-5 Grand-Slam Match	simulation
2-6 Chasing the Circle	geometrical probability
2-7 Estimating π	geometrical probability
Extending	
2-1 Bazuka Bats	simulation
2-2 Rock-Star Cards	simulation
2-3 Raining in Surfers' Paradise	simulation

Calculator Helps

For Module 2:

Calculator Help 5: Using the Random-Number Generator

CONTENT BACKGROUND

Experimental and Theoretical Probabilities

Activities or experiments involving chance result in outcomes that normally cannot be predicted with certainty. The study of probability aims to establish numerical values that convey the likelihood of a particular outcome or event. For example, the data in Figure 1 present the free-throw shooting record of the top five players on the Newtown State University Women's Basketball Team, the Bluebirds.

Figure 1
Free-Throw Shooting
Record:
1994 Newtown University
Women's Basketball Team

Player	FT Attempts	FT Percentage
Laura	$\frac{30}{50}$	.600
Melissa	$\frac{57}{72}$	.792
Stacy	$\frac{42}{75}$	.560
Arantxa	$\frac{27}{43}$	.628
Mia	$\frac{20}{31}$	.645

To interpret these figures we need a context. Suppose Laura is at the free-throw line for a single shot. We know that her free throw will result in one of two outcomes: She makes the shot or she misses the shot. The outcomes are said to be the *sample space* for the activity of attempting a free throw.

From the data available, the probability that Laura will make the shot is 0.60 because 60 percent of her free-throw shots have been successful. In this case we are using experimental probability, defined as follows:

$$\text{experimental probability} = \frac{\text{number of successes}}{\text{number of trials}}$$

Experimental probability is determined from existing data, as in this example, or by carrying out a number of trials of an experiment and recording the number of times a desired event occurs. Each occurrence of an event is counted as a success.

By way of contrast, theoretical probabilities are determined by physical characteristics, geometrical properties, or by other probability calculations. For example, when flipping a coin we assign a probability of 0.50 to a head and also to a tail because the coin is symmetrical and heads and tails are equally weighted. Likewise, each of the numbers 1 through 6 is assigned a probability of $\frac{1}{6}$ when a fair die is rolled.

Outcomes are not always equally likely. For example, at a show for baseball card collectors, if we draw from a mystery box that has two star-player cards and ten common-player cards, the probability of getting a star-player card is $\frac{2}{12}$, and the probability of getting a common-player card is $\frac{10}{12}$.

Fundamental Properties of Experimental and Theoretical Probabilities

Both experimental probabilities and theoretical probabilities have the following properties:

- The probability of an event is a real number between 0 and 1 inclusive.
- The sum of the probabilities of all outcomes in the sample space is 1. See references [5], [9], and [11].

To illustrate let us return to Laura at the free-throw line. There are two outcomes: makes the shot and misses the shot. Because she has made $\frac{30}{50}$ free throws during the season she must have failed to make $\frac{20}{50}$ free throws. Her probabilities are: 0.60 for a "make" and 0.40 for a "miss." Both of these probabilities are between 0 and 1 and because these are the only two outcomes, the sum of the probabilities is 1. We observe that the event "either makes the shot or misses the shot" is certain to occur; this event has the maximum allowable probability of 1. By way of contrast it is impossible for the ball to stick on the rim, and this event has the minimum possible probability of 0.

PROBABILITIES AND ODDS

Based on a record of 15 wins and 5 losses, the probability of the Newtown State University Bluebirds winning a game is 0.75. Given this, we also know that the probability of the Bluebirds losing is 0.25. In this case, the event "losing a game" is called the *complement* of the event "winning a game." As is always true of complementary events, the probability of winning and the probability of losing sum to 1.

When the probability of winning a game is 0.75, this means that over a span of many games the Bluebirds can expect to win three and lose one of every four games. In essence, the ratio of their chances of winning to losing is 3:1. This ratio is said to be the odds in favor of a Bluebird victory. By way of contrast, the odds against a Bluebird victory are 1:3.

In general, the odds in favor of an event is defined as follows:

$$\text{odds in favor} = \frac{\text{probability of the event}}{\text{probability of its complement}}$$

From this definition we observe that the odds in favor of the Bluebirds winning is $\frac{3/4}{1/4} = \frac{3}{1}$. This is in agreement with our intuitive considerations shown above.

The odds against an event is similarly defined:

$$\text{odds against} = \frac{\text{probability of the event's complement}}{\text{probability of the event}}$$

Again, we observe that the odds against the Bluebirds winning is $\frac{1/4}{3/4} = \frac{1}{3}$. This is consistent with our earlier discussion. See reference [9].

Using Tree Diagrams to Determine the Theoretical Probability of Two-Stage Situations

Even though the emphasis in the students' activities is on modeling and simulation, you may wish to devote some time to determining theoretical probabilities. It also will be helpful to use theoretical probabilities to gauge the accuracy of the student simulations.

PROBLEM 1

Melissa is at the foul line for a "two-shot" penalty. What is the theoretical probability that she makes both shots?

SOLUTION

The probability that Melissa makes a good shot is 0.792. We signify this by P(G) = 0.792. We also note that the probability that she does not make the shot is 0.208. This is denoted by P(N) = 0.208.

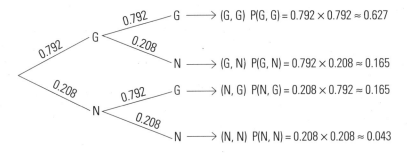

G represents "makes the shot"
N represents "misses the shot"

The tree diagram identifies the four two-stage outcomes and their respective probabilities. If A is the event that Melissa makes both shots, A = {(G,G)}, then P(A) = 0.627.

PROBLEM 2

Later in the same game Melissa is at the line for a "one-and-one." What is the probability that she scores at least one point?

SOLUTION

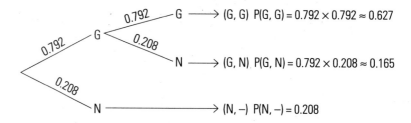

G represents "makes the shot"
N represents "misses the shot"

The tree diagram shows the three outcomes with their respective probabilities. Note that when Melissa misses the first shot there is no second shot. As a result the set of simple outcomes in the sample space is {(G,G); (G,N), (N,–)}, where the dash indicates there was no second shot. Even though the outcomes are not all two-stage outcomes, the tree diagram shows that the sum of the probabilities of the simple outcomes is equal to 1.

Let B be the event that Melissa scores at least one point. Then: B = {(G,G); (G,N)}. P(B) = 0.627+ 0.165 = 0.792.

NOTES

1. It is assumed that all events are independent, that is, the result of Melissa's second shot is independent of the outcome of her first shot.
2. When two events A and B are independent, $P(A \cap B) = P(A) * P(B)$
3. When two events A and B are not independent, $P(A \cap B) = P(A) * P(B|A)$ and $P(A \cap B) = P(B) * P(A|B)$, where $P(B|A)$ is the conditional probability of B given A. $P(A|B)$ is defined in a similar way.
4. The results described in notes 2 and 3 provide the basis for multiplying probabilities along the branches of a tree diagram.
5. References: [9] and [11].

SIMULATION

Simulations are widely used in science, business, government, and industry to determine the likelihood that some event will occur under conditions equivalent to the real situation, but that are far less costly in time, money, or life. The testing of aircraft in wind tunnels and pilots' use of flight simulators are two examples of simulated experiments. In probability situations, simulation helps students obtain the experimental probability of some real-world event. Students carry out an experiment using a probability device that models the real-world situation.

Because the long-run experimental probability approximates the theoretical probability, simulation can lead to valid estimates of the required probability. Simulation also is a powerful problem-solving tool in its own right that can be used by students at the middle-school level and beyond.

The following demonstrates how the process of simulation can be used to solve complex probability problems, the theoretical solutions of which are well beyond the capabilities of most middle-school and high-school students. In describing the simulations, this guide uses a procedure similar to that presented by the authors of references [10] (pages 3–4) and [11] (pages 80–87). A description of the use of probability devices that generate random outcomes precedes the simulation problems.

GENERATING RANDOM OUTCOMES

Suppose Bonnie, another player on the Newtown State University Women's Basketball Team, has a free-throw shooting record of 0.50. When Bonnie steps to the free-throw line, there are two possible outcomes: "makes the shot" with probability 0.50 and "misses the shot" with probability 0.50.

The toss of a coin perfectly models Bonnie's shooting, because the coin has two outcomes each with probability 0.50. The model is complete when we designate heads as "makes the shot" and tails as "misses the shot." So long as Bonnie's shooting probability remains at 0.50, the coin continues to simulate her free-throw shooting.

Alternatively, you could designate the outcomes of rolling a die to model Bonnie's shooting. Because there are six equally likely outcomes when a fair die is rolled, the even numbers could represent "makes the shot" while the odd numbers could represent "misses the shot." Again, there are two events, evens or odds, and the probability of each event is 0.50. You could also use other groupings of die rolls, (1, 2, 3) and (4, 5, 6) for example, to represent the two outcomes.

Finally, you could use a spinner divided into two regions of equal area, one region representing "makes the shot" and the other representing "misses the shot." Each spin of the spinner then simulates one of Bonnie's free-throw attempts.

Probability devices such as these have physical limitations. For example, Stacy has a probability of 0.56 of making a free throw. Neither a coin nor a die could be used to model Stacy's free-throw shooting. Moreover, it is very difficult to accurately divide a spinner so that one region represents 56 percent and the other represents 44 percent. Randomly generated numbers do not have such physical limitations and are capable of simulating even the most complex problems.

Using Random Numbers

The TI graphics calculators generate sets of ten-digit rational numbers between 0 and 1. To access the random-number generator, press $\boxed{\text{MATH}}$, select the MATH PRB menu by pressing the $\boxed{>}$ arrow key, select 1:rand and then press $\boxed{\text{ENTER}}$. A random number appears on the screen as in Figure 2(a). Repeatedly pressing $\boxed{\text{ENTER}}$ produces additional ten-digit rational numbers. Figure 2(b) shows a typical screen.

Figure 2
Random Numbers

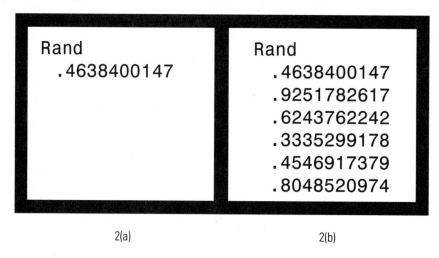

2(a)   2(b)

In using the random-number generator for simulation, use single digits or groups of digits with no regard for the decimal point displayed on the screen. The random display in Figure 2(b) can be used to simulate Bonnie's free-throw shooting if we agree that the single digits 0 through 4 (that is, 5 of the 10 possible digits) represent "makes the shot" and the single digits 5 through 9 (also 5 of the 10 possible digits) represent "misses the shot." Hence, the random digits in Figure 2(b) can be translated into "makes the shot" or, "misses the shot" as is shown in Figure 3. This assures that the outcomes are being assigned in such a way that the probability of makes the shot is 0.50 and the probability of misses the shot is 0.50.

Figure 3
Translations of
Random Digits

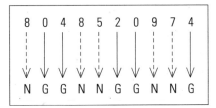

This is an example of how random numbers can be grouped to perfectly model probability situations. Moreover, random numbers have great modeling versatility that will be shown in the simulations that follow.

Simulation 1

Use random numbers to simulate ten free-throw attempts by Laura. How many times did she make the shot? The following is the simulation sequence.

Step 1: Model
There are only two outcomes: "makes the shot" and "misses the shot." Because the probability that Laura makes the shot is 0.60, the probability she misses is 0.40.

Use single-digit random numbers to model Laura's shooting. Make the following assignment between a random number and whether Laura makes or misses the shot:

digits 0, 1, 2, 3, 4, 5 = "makes the shot"
digits 6, 7, 8, 9 = "misses the shot"

This is a valid model because each single-digit number will represent a "make" or a "miss" with theoretical probability 0.60 (6 out of 10) and 0.40 (4 out of 10), respectively.

Step 2: Trial
A trial consists of generating a random digit and noting whether it is in the set 0 through 5 ("make") or 6 through 9 ("miss").

Step 3: Repeat
Ten trials are required to simulate ten free throws.

Step 4: Record the Observation of Interest
Ten single-digit random numbers from a TI calculator are shown on page 28.

STEP 5: SUMMARIZE AND DRAW CONCLUSIONS

In the ten random numbers there are four in the set 0 through 5 and six in the set 6 through 9. In these ten trials Laura makes the shot four times and misses it six times.

NOTES

1. Other number groupings can be used to model Laura's shooting. Here is an example:

    digits 4, 5, 6, 7, 8, 9 = "makes the shot"

    digits 0, 1, 2, 3 = "misses the shot"

    The probabilities remain 0.60 for a "make" and 0.40 for a "miss."

2. With probabilities given in tenths, it was possible to model with single-digit random numbers. What if the probabilities had been 0.56 for "makes the shot" and 0.44 for "misses the shot"?

3. The free-throw shooting can be modeled in other ways. For example, you can use a spinner, such as the one shown here, that has ten regions of equal area.

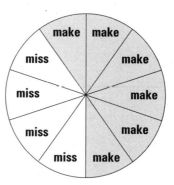

4. For further reference on simulations, see references [10] and [11].

Simulation 2

Use random numbers to simulate ten free-throw attempts by Stacy. How many times did she make the shot? The following is the simulation sequence.

STEP 1: MODEL

There are only two outcomes: "makes the shot" and "misses the shot." Because the probability that Stacy makes the shot is 0.56, the probability she misses is 0.44.

To model Stacy's shooting we must use two-digit random numbers. Make the following assignment between a random number and whether Stacy makes or misses the shot:

00, 01, 02, . . ., 54, 55 = "makes the shot"
56, 57, 58, . . ., 98, 99 = "misses the shot"

This is a valid model because each two-digit number will represent a "make" or a "miss" with theoretical probability 0.56 (56 out of 100) and 0.44 (44 out of 100), respectively.

STEP 2: TRIAL

A trial consists of generating a two-digit random number and noting whether it is in the set 00 through 55 ("make") or 56 through 99 ("miss"). With the TI calculators a trial is formed by grouping two single-digit random numbers into a pair, thus creating a two-digit random number from the set 00 through 99.

STEP 3: REPEAT

Ten trials are required to simulate ten free throws.

STEP 4: RECORD THE OBSERVATION OF INTEREST

Random numbers from a TI calculator are shown here with each group of two digits identified.

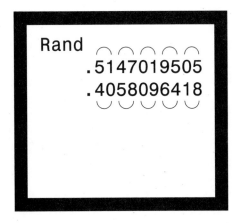

This results in the following ten two-digit random numbers:
51, 47, 01, 95, 05, 40, 58, 09, 64, 18

Step 5: Summarize and Draw Conclusions
In the ten two-digit random numbers, there are seven in the set 00 through 55 and three in the set 56 through 99. In these ten trials, Laura makes the shot seven times and misses it three times.

Notes
1. Simulations 1 and 2 illustrate the use of TI-calculator-generated random numbers to represent one- and two-digit random numbers. How can the TI calculator be used to model free-throw attempts if the probability of "makes the shot" is 0.628?
2. Simulation 1 Note 3 identifies another way to generate random outcomes to simulate Laura's free-throw shooting. By what other means, aside from random numbers, could you generate random outcomes to simulate Stacy's free-throw shooting? Compare the effectiveness and efficiency of the alternatives with the use of random numbers.

Simulation 3

Laura is at the free-throw line for two shots. What is the probability that she will make both the shots? The following is the simulation sequence.

Step 1: Model
There are only two outcomes for each shot: "makes the shot" and "misses the shot." For Laura at the line you will use the same assignments as in Simulation 1:

> digits 0, 1, 2, 3, 4, 5 = "makes the shot"
> digits 6, 7, 8, 9 = "misses the shot"

Step 2: Trial
Because Laura takes two shots, one trial consists of generating a pair of single-digit random numbers and noting whether both are in the set 0 through 5 ("make") or not.

Step 3: Repeat
You need a number of trials from which you will determine experimental probability. Consider the results from 30 trials.

Step 4: Record the Observation of Interest
You are interested in successes (Laura makes both shots) and non-successes (Laura does not make both shots). Record whether a trial is a success.

Step 5: Summarize and Draw Conclusions
A TI calculator screen is shown on page 31.

```
Rand
    .7223793157
    .0125655621
    .4211182917
    .3076351501
    .9722077719
    .0288170136
```

It contains 60 single-digit random numbers that produce the 30 trials shown below:

7-2	**2-3**	7-9	**3-1**	5-7
0-1	**2-5**	6-5	5-6	**2-1**
4-2	**1-1**	1-8	2-9	1-7
3-0	7-6	**3-5**	**1-5**	**0-1**
9-7	**2-2**	0-7	7-7	1-9
0-2	8-8	1-7	**0-1**	3-6

The first trial, 7-2, represents a "miss" on the first attempt (7 is from the set 6 through 9) and a "make" on the second attempt (2 is from the set 0 through 5). This is not a success. The second trial, 2-3, represents two shots made, and is, therefore, a success.

Successes are shown in bold text above. There are 14 successes in 30 trials. The experimental probability of success is $\frac{14}{30}$ or approximately 0.467. Therefore, the probability that Laura makes both shots is 0.467.

NOTES

1. The theoretical probability that Laura makes both shots, assuming the two shots are independent of each other, is $0.60 * 0.60 = 0.36$.
2. It is important for students to distinguish between each single-digit random number used to model one of Laura's shots and the pairing of random digits required to identify a trial.

Simulation 4

Stacy is at the free-throw line for a one-and-one foul shot. What is the probability she will score two points? The following is the simulation sequence.

STEP 1: MODEL
Use the same model as in Simulation 2:

00, 01, 02, . . . , 54, 55 = "makes the shot"
56, 57, 58, . . . , 98, 9 = "misses the shot"

STEP 2: TRIAL
Because Stacy is at the line for a one-and-one opportunity, a trial will require one two-digit random number if she misses the first shot, but will require two two-digit random numbers if she makes the first shot. To complete a trial generate a two-digit random number and note whether it represents a "make" or a "miss." If the first shot is a "miss" the trial is complete. If Stacy makes the first shot consider the next two-digit random number and note whether it represents a "make" or a "miss."

STEP 3: REPEAT
You need a number of trials from which you will determine experimental probability. Consider the results from 20 trials.

STEP 4: RECORD THE OBSERVATION OF INTEREST
You are interested in successes (Stacy makes both shots) and non-successes (Stacy misses the first shot or misses the second shot). Record whether a trial is a success.

STEP 5: SUMMARIZE AND DRAW CONCLUSIONS
A TI calculator screen is shown here.

```
Rand
 .7245328176
 .3017450175
 .2492764264
 .9461839774
 .9252198295
 .0427909441
```

It contains 60 single-digit random numbers. Here are digit pairs to represent individual free throws:

72	45	32	81	76
30	17	45	01	75
24	92	76	42	64
94	61	83	97	74
92	52	19	82	95
04	27	90	94	41

The first two-digit number, 72, represents a "miss" on the first shot of a one-on-one. A "miss" signals an end to a trial. Move to the next number, 45, to begin another trial. The number 45 represents a "make" on the first attempt of the one-on-one. Therefore, look to the next number, 32, to complete the trial. The 32 represents a "make" on the second shot, and thus you have a success—two consecutive shots made by Stacy. The numbers below show the same random numbers grouped for the first 20 trials.

72 **45-32** 81 76 **30-17** **45-01** 75   24-92 76 42-64
94 61   83 97 74   92   **52-19** 82   95 **04-27**

Successes are shown in bold text. There are 5 successes in 20 trials. The experimental probability of success is $\frac{5}{20}$ or 0.25. Therefore, the probability that Stacy scores two points is 0.25.

NOTES
1. The theoretical probability that Stacy will score two points on the one-on-one free-throw opportunity is 0.31.

Simulation 5

Nancy is another member of the Newtown State University Women's Basketball Team. She has an interesting free-throw shooting record this season: If she makes a free throw, the probability she will make the next free throw is 0.7. If she misses a free throw, the probability she will miss her next shot is 0.6.

Carry out a simulation to determine the mean number of free throws Nancy will make in a game where she attempts five free throws. Assume that she always makes her first free-throw attempt of a game. The following is the simulation sequence.

STEP 1: MODEL
Use the notation $P(G|G)$ to represent the probability of Nancy *making her next free throw given that she just made a free throw* and use $P(N|N)$ to represent the probability of Nancy *missing her next free throw given that she just missed a free throw.* From the problem statement, you have $P(G|G) = 0.70$ and $P(N|N) = 0.60$. Using complements you can determine $P(N|G)$, the probability of missing the next free throw given that Nancy just made a free throw, and $P(G|N)$, the probability of making the next free throw given that she just missed a free throw. You have $P(N|G) = 1 - 0.7 = 0.3$ and $P(G|N) = 1 - 0.6 = 0.4$.

Use single-digit random numbers to model Nancy's free-throw shooting, as shown in the table below.

Random Number Assignment	Nancy makes her next free throw	Nancy misses her next free throw
Nancy just made a free throw	0 1 2 3 4 5 6	7 8 9
Nancy just missed a free throw	0 1 2 3	4 5 6 7 8 9

STEP 2: TRIAL

Assuming Nancy's first free-throw attempt is good, a trial consists of generating four single-digit random numbers and, depending on the result of the previous shot and the conditional probabilities, noting whether each shot is a "make" or a "miss." (See the table above.)

STEP 3: REPEAT

You need a number of trials from which you will determine the desired mean value. Consider the results from 10 trials.

STEP 4: RECORD THE OBSERVATION OF INTEREST

You are interested in the number of free throws made in five attempts, assuming that the first free throw is made each time. For each trial, record a number, 1 through 5, to represent the number of free throws Nancy made in a game.

STEP 5: SUMMARIZE AND DRAW CONCLUSIONS

A TI calculator screen is shown here.

```
Rand
    .6278969608
    .6174036363
    .8068418276
    .7734326792
```

It contains 40 single-digit random numbers. This gives you 10 sets of 4 numbers:

6-2-7-8	9-6-9-6	0-8-6-1	7-4-0-3	6-3-6-3
8-0-6-8	4-1-8-2	7-6-7-7	3-4-3-2	6-7-9-2

Assuming Nancy makes her first free throw in each game, the groups of random numbers translate into the following record of "makes" (G) and "misses" (N) for Nancy:

G-GGNN	G-NNNN	G-GNNG	G-NNGG	G-GGGG
G-NGGN	G-GGNG	G-NNNN	G-GGGG	G-GNNG

The result of each trial was determined by assuming Nancy made her first free throw (G) and then referring to the random-number assignments shown in the table on the previous page. Then you record these numbers to represent the free throws made in each trial:

3	1	3	3	5
3	4	1	5	3

The mean number of free throws is 3.1 (31 free throws made in 10 games).

NOTES

1. The assignment of random numbers must be in accordance with each conditional probability. For example, the assignment for $P(G|G)$ is different from $P(G|N)$.
2. The result of each trial must be carefully recorded. Each outcome that is part of a trial, and the relevant probability of that outcome, depends on the previous outcome.
3. Consider these additional questions in the context of this problem:

   • What is the longest run of made free throws in the simulation? What is the probability of that length run, based on the simulation?

   • If we assume Nancy misses her first free-throw attempt, will the mean number of made free throws per game change? Perform a simulation to verify your conjecture.

CLASSROOM ACTIVITIES

The following activities may be useful in examining a number of the probability concepts considered in Module 2.

Activity 1—Heads and Tails

At the start of this game all students stand and predict how two coins will land when you flip them.

   • Have students place both hands on their heads to predict *two heads.*

   • Have students place one hand on their heads and one hand on their tails to predict *one of each.*

   • Have students place both hands on their tails to predict *two tails.*

After each flip of the two coins, only the students predicting correctly remain standing. Those still standing then make a new prediction about the next flip, and the game continues until only the winning student remains standing. Note that students can change their prediction before each flip of the two coins.

The class might discuss which of the three outcomes is the best prediction and whether or not there is a "best" strategy.

Activity 2—Dice Game

In this game a die is rolled and students are given the opportunity:

- to get 5 points or 6 points whenever 5 or 6 is rolled on the die or
- to get 1, 2, 3, or 4 points whenever one of these numbers is rolled.

The first player to get 20 points wins the game. After playing the game, the class might discuss which choice offers the best chance of a win and why.

Activity 3—Birthday Problem

To begin this activity, invite students to predict the probability that at least two people in a randomly formed group of five people have the same birth month. Record student predictions in some appropriate way (for example, those who think there is much more than a 50 percent chance, those who think there is about a 50 percent chance, and those who think there is much less than a 50 percent chance).

Have students randomly form groups of five. Instruct students to determine whether at least two people in their group have the same birth *month* (not necessarily the same day or year). Use the results to find the experimental probability that, in a group of five people, at least two people have the same birth month. Compare this with students' initial predictions.

The class might discuss why the probability is higher than expected.

Activity 4—Let's Make a Deal

To start the activity, describe the following situation:

During a certain game show, a contestant is shown three closed doors. One of the doors has a very expensive prize behind it, and each of the other doors has a worthless prize. The contestant is asked to pick a door. The game show host opens one of the remaining closed doors and shows it to the contestant—always revealing a worthless prize. The contestant is then given the option to stick with the original choice or to switch to the other unopened door. What should the contestant do?

Have students vote on whether it is better to *stick* or *switch* or whether it doesn't matter. Then invite students to simulate the situation in order to determine the experimental probabilities for stick and switch. In carrying out the simulation, students must first collect data using only the *stick* strategy and then collect data using only the *switch* strategy (or vice versa).

For example, each group of students might use three envelopes to represent the three doors. Two envelopes contain a card naming a gag gift, and the other names an expensive prize. One student acts as host, a second students acts as a contestant always using the *stick* strategy, and a third student records the outcomes. The students then change roles to record data for the *switch* strategy. After the simulation, students might discuss why the *switch* strategy has a probability twice that of the *stick* strategy. Intuitively, the students might reason as follows: The *stick* strategy has a probability of $\frac{1}{3}$ because there are three possible doors, and only one has the expensive prize. On the other hand, the *switch* strategy has a probability of $\frac{2}{3}$ because a wrong choice can initially be made in two out of three ways, and the host takes care of one of these.

IMPLEMENTATION GUIDELINES

Teaching Suggestions by Activity

Starting 2-1 Flip the Coin An alternative way of looking at heads and tails is to balance a nickel on its side, hit the table, and then record "head" or "tail." Under these conditions, more heads can be expected. Students might discuss why this is so (for example, the nickel is heavier on the one side).

Starting 2-2 Model the Coin Flip Discuss the distinction between experimental and theoretical probability and the meaning of equally likely events (see *Content Background* on page 20). Note that this is the first activity to use random numbers to model an experiment.

Starting 2-3 Modeling Two Flips This activity examines what happens on two consecutive flips of a coin. Random numbers are used to model this situation and this requires grouping single-digit random numbers in pairs (see *Content Background* on page 30).

Starting 2-4 Steffi's Serve If students are unfamiliar with tennis, discuss the term *double fault*. A double fault occurs when two bad serves occur in a row. While a tennis player is allowed two serves, there is no second serve if the first serve is good. Assignment of random numbers can be adjusted accordingly.

Starting 2-5 Flip the Cup Students need to come to a consensus about what will be accepted as ''up'' versus ''down'' when flipping the cup. Have students check the sum of their probabilities at the end of their experiment to note that this sum equals one. Paper cups with different dimensions will produce different results. You could build an assessment activity based on determining and analyzing probabilities associated with paper cups having different dimensions.

Starting 2-6 Chances for the Name In preparation for this activity, it may be appropriate to review *histograms* (see Appendix E, ''Word Bank,'' in the Student Edition). Students use a histogram to find the number of students whose first names have four consonants. They then use this to find the experimental probability that a name drawn at random will have four consonants.

Starting 2-7 Modeling the Chances for the Name Students may want to use a calculator to determine the central angles of the circle graph used to construct the spinner.

Starting 2-8 Two Broadway Favorites You may wish to review some basic ideas of probability before students undertake this activity. See *Content Background* on pages 20–24.

Starting 2-9 Three in a Row Because the probability that a student has seen both shows is 0.6, students will be able to use single-digit numbers to carry out the simulation. They will need help in recognizing that random numbers need to be grouped in threes in order to consider probabilities associated with three consecutive draws.

Developing 2-1 Three Hits Batting averages do not stay constant—they tend to change each time a player comes to bat, depending on whether or not a hit is made. In this activity, assume that the batting average remains constant. This problem will be reasonably complex, because a batting average of .325 will require three digits to model its probability. It will also be necessary to generate about 30 groups of three "at bats" for this experiment, which means nine digits per trial.

Developing 2-2 The Top Ten and Developing 2-3 On Top for Four Weeks These activities are similar to Developing 2-1. However, the term *at most* is used, and students need to understand that this means "less than or equal to." In the same way, it is a good opportunity to consider *at least,* which means "greater than or equal to."

Developing 2-4 Basketball Playoffs Students should consider how the series is played and note that one team needs to win four games or the other lose four in order to determine a winner. If possible, encourage students to compare their experimental probabilities with the theoretical probability.

Developing 2-5 Grand-Slam Match Note that three wins or three losses by a player determines the end of a grand slam match. Students also need to understand "odds" and the use of complements as both ideas occur in this activity (see Appendix E, "Word Bank," in the Student Edition and *Content Background* on pages 22–23).

Developing 2-6 Chasing the Circle In this activity you might like to find a class mean for the number of circles pricked. You could use the overhead TI graphics calculator to enter student data as they are reported. To find the mean use <1-Var> statistics under the STAT CALC menu.

Developing 2-7 Estimating π Students should measure the diameter of the circle in millimeters. Poll the class to get an average estimate for this measure.

Extending 2-1 Bazuka Bats Students need to take care in assigning random numbers to determine "0," "at least 2," and "4 hits."

Extending 2-2 Rock-Star Cards This activity could be carried out using a die rather than random numbers. One of the numbers will need to be discarded because there are five rock star cards.

Extending 2-3 Raining in Surfers' Paradise Because Sunday is sunny, students only need to simulate six days. As probabilities change, students need to reassign random numbers. This activity could be carried out using a spinner rather than random numbers. (See *Content Background* on pages 33–35.)

MODULE 3:
Counting

OVERVIEW AND PURPOSE

The study of counting is fundamental not only to problems involving probability but also to problems that involve both ordered and nonordered arrangements and selections. Today, counting principles are used extensively in coding credit cards, generating telephone numbers, scheduling work assignments and sporting rosters, and determining traffic routes. In this module students will be exposed to counting problems from a variety of real-world settings and will also use counting techniques to solve more complex probability problems. Students will learn how to:

- recognize when to apply addition and multiplication operations to counting problems;
- use the *pigeonhole principle* to solve problems that guarantee certain requirements;
- apply counting principles associated with permutations and combinations; and
- explain the relationship between permutation and combination counting.

For much of this module students solve problems designed to enhance their understandings of more advanced counting techniques. The role of the TI graphics calculator in this module is essentially a supportive one. It enables students to explore their own intuitive counting processes, investigate relationships, check their solutions, and solve problems involving large numbers.

Although the conceptual understanding of counting is substantially developed through the activities in Module 3, student learning will be facilitated if you provide some initial development of the key counting processes.

With this in mind the following sections provide:

- an outline of key mathematical ideas by activity;
- content background;
- sample classroom activities; and
- guidelines for implementing the activities of Module 3.

OUTLINE OF KEY MATHEMATICAL IDEAS BY ACTIVITY

Activity Title		Key Mathematical Ideas
Starting		
3-1	Naming the Dog	addition rule
3-2	How Many Faces?	addition, multiplication rules
3-3	Birth Month in a Box	pigeonhole
3-4	Three the Same	pigeonhole
3-5	First Names in a Box	pigeonhole
3-6	How Many Sandwiches?	multiplication rule
3-7	How Many Kinds of Pizza?	multiplication rule
Developing		
3-1	Batting Order	factorial
3-2	Pitcher Bats Last	factorial with restriction
3-3	Raffle Tickets	permutations
3-4	Calculating the Number of Raffle Tickets	permutations
3-5	"Special" Raffle Tickets	permutations
3-6	"Extra-Special" Raffle Tickets	permutations
3-7	Mini-Lotto	combinations
3-8	Calculating the Mini-Lotto Pairs	combinations
3-9	Pick Three	combinations
3-10	Pick Three Again	combinations
Extending		
3-1	Many the Same	counting
3-2	Two in the Same Month	counting
3-3	The State Lottery	counting
3-4	Winning Hands	counting

CONTENT BACKGROUND

The following problems (on pages 43–51) deal with three fundamental principles of counting—the addition principle, the multiplication principle, and the pigeonhole principle. Each of these techniques will be developed through reference to relevant problems drawn from a restaurant setting. Permutations and combinations are also examined. (See references [2] and [3].)

Figure 1 displays menu information from Blaise's Bistro.

Figure 1

Blaise's Bistro Menu

A-LA-CARTE	PIZZA	BEVERAGES
Soup:	Size:	Soda:
Vegetable	Large	Cola
Onion	Medium	Diet Cola
	Small	Lemonade
Entree:		Ginger Ale
	Crust:	Root Beer
Meat	Deep dish	Diet Root Beer
Chicken Kiev	Traditional	
Prime rib		Other:
	Toppings:	Coffee—regular
Greens	Sausage	Coffee—decaf
Spinach	Pepperoni	Tea—hot
Peas	Anchovies	Iced Tea
Beans	Beef	Milk
Broccoli	Ham	
Potatoes		
Cheese		
Bacon		
Sour Cream		
Chives		
Butter		**SPECIALS**
		10% OFF
Dessert:		
Key Lime Pie		•
Apple Pie a la mode		•
Black Forest Cake		•
French Vanilla Ice		
Cream		

The Addition Principle

PROBLEM 1

On Tuesdays, Blaise's Bistro offers only a choice from the Greens or Potatoes for a vegetable, but not both. How many different choices of vegetable can be made on Tuesdays?

SOLUTION

Figure 2 shows one way to represent the two sets that must be considered. Note that the two sets have no elements in common; that is, they are *disjoint* sets.

Figure 2

One Way to Represent
the Two Sets

Greens	Potatoes
Spinach	Cheese
Peas	Bacon
Beans	Sour Cream
Broccoli	Chives
	Butter

Because a customer may choose from four kinds of green vegetables or from five kinds of potatoes, nine distinct choices are
possible. The number of choices is the sum of the number of
elements in the disjoint sets.

COMMENTS

In this problem the *addition principle* is applied to a situation with
two disjoint sets having four elements and five elements, respectively. This principle can be extended to a more general situation.
(See references [2] and [3].)

Suppose there are a number of sets, say five, with a elements in
the first set, b elements in the second set, c elements in the third
set, d elements in the fourth set, and e elements in the fifth set. If
the sets are disjoint, then the total number of elements in the sets
is $a + b + c + d + e$. (Note that the choice of five sets is arbitrary,
for the principle holds for all counting numbers.)

The Multiplication Principle

PROBLEM 2

Blaise's Bistro pizza menu includes a choice of one of three sizes,
two types of crust, and a choice of one of five toppings. How many
different varieties of pizza could be purchased?

SOLUTION

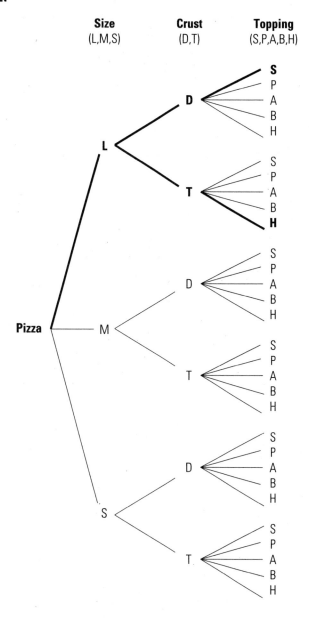

The tree diagram here illustrates the varieties of pizza that can be ordered. For each size pizza there are two different crusts, and for each type of crust there are five different toppings to choose from. In the case of a large pizza, there will be 2 * 5 = 10 varieties, two of which include a large deep-dish sausage pizza and a large thin-crust ham pizza. These two examples are highlighted in the tree diagram. With three possible sizes, there are 3 * 10 = 30 overall varieties.

COMMENTS

In the problem above, the *multiplication principle* is used in a situation where three sequential choices have to be made. The first choice (pizza size) can be made in three ways, the second choice (crust type) in two ways, and the third choice (topping) in five ways. Under these conditions, the branches of the tree diagram demonstrate that the number of different varieties is 3 * 2 * 5.

In general, suppose that a procedure can be divided into a sequence of five stages and the first stage can be performed in a ways, the second in b ways, the third in c ways, the fourth in d ways, and the fifth in e ways. Then the number of different ways the entire procedure can be performed is $a * b * c * d * e$. Note that the use of five stages is arbitrary. The multiplication principle can be defined for any natural number of stages. Moreover, in the case of the multiplication principle, the choice made at any stage does not depend on the particular choice made at a previous stage. (See references [2] and [3].)

The Pigeonhole Principle

PROBLEM 3 PART (A)

Blaise's Bistro offers six different sodas. A group of 30 people comes into the bistro. How many members of the group must purchase soda so that at least two purchase the same soda?

SOLUTION

In the worst possible scenario, the first six members will all order a different soda. In this case, the seventh member must purchase a soda that has already been purchased. Hence, seven members must purchase sodas to ensure that at least two purchase the same soda. The possibility of a duplication occurring before the seventh person, for example the first two members purchase a diet cola, is not relevant in this problem.

COMMENTS

The *pigeonhole principle* is used to solve this problem and Problem 3(b). This principle states that if a set of pigeons is placed into pigeonholes and there are more pigeons than pigeonholes, then some pigeonholes must contain *at least* two pigeons. (See references [2] and [3].) The diagram on page 47 illustrates the use of the pigeonhole principle in Problem 3(a).

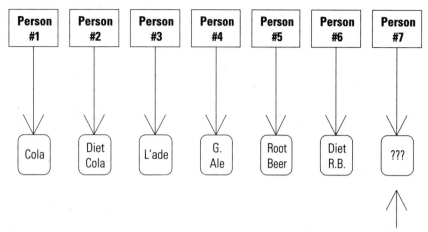

Person #7 is **forced** to select a beverage previously selected.

PROBLEM 3 PART (B)

How many members of the group must purchase a soda in order to ensure that at least three people purchase the same soda? How many are needed to ensure that at least four purchase the same soda?

SOLUTION

The solution to problem 3(a) tells us that there will be at least one duplication after seven members have purchased a soda. Continuing a worst scenario situation, each of the six sodas will be ordered by 2 people when 12 people have made their purchase. Hence the 13th member will have to select the same soda that 2 members have already purchased. With a similar argument 19 members will need to purchase sodas to be certain that 4 members get the same soda. Notice the pattern that develops:

At least 2	7
At least 3	13
At least 4	19
•	•
•	•
•	•
At least n	$6n - 5$

Permutations and Combinations

PROBLEM 4 PART (A)

Blaise's Bistro sometimes varies the order of the soda beverages on the menu board. How many different ordered arrangements are possible for the six sodas?

SOLUTION

This problem exemplifies the multiplication principle. With no restriction on placement, the first drink can be selected in six ways, the second in five ways, the third in four ways, and so on. Hence, the number of ordered arrangements equals
6 * 5 * 4 * 3 * 2 * 1 = 720.

Products such as 1 * 2 * 3 * 4 * 5 * 6 are expressed in a special form called *factorial notation*. In this case, the product can be written 6!. In general, 1 * 2 * . . . * n = n!. On the TI-82 calculator this is found by pressing the [MATH] button, using [<] to move to the PRB menu, and selecting **4:!**. On the TI-81 calculator it is found by pressing the [MATH] button, leaving the screen on the [MATH] menu, and selecting **5:!**.

PROBLEM 4 PART (B)

In problem 4(a) Blaise's Bistro placed no restrictions on the placement of any soda. If each diet soda must be placed next to its corresponding non-diet soda (such as diet cola and cola together), how many orderings are possible under this restriction?

SOLUTION

In this case, we think of cola and diet cola as being joined; likewise for root beer and diet root beer, as shown in the figure.

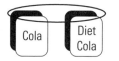 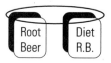

There are now four spaces to be filled on the menu board, two of which are doubles. By the multiplication principle, these spaces can be filled in 4! ways. However, this does not conclude the solution, because each of the joined sodas (regular-diet) can be arranged in two ways, namely regular-diet or diet-regular. Again using the multiplication principle, the total number of ordered arrangements is 2 * 2 * 4! = 96. Students may benefit from listing some of these cases, especially those involving the reversal of the diet-regular pairs.

PROBLEM 5

In yet another promotional activity, Blaise's Bistro offers three sodas at a special discount rate of 10 percent off. Each day he writes these three specials in the bottom right corner of his menu board. How many different orderings of three sodas are possible?

SOLUTION

There are three vertical places to fill, hence the sodas can be read in various orders. The first place can be filled in six ways, because any of the six sodas can be on special and can be in the first place. The second place can be filled in five ways, and the third place in four ways. Using the multiplication principle, the number of ordered arrangements of three sodas chosen from six sodas is $6 * 5 * 4 = 120$. This is said to be a permutation of six sodas taken three at a time.

COMMENTS

A *permutation* is an arrangement of objects in a definite order where none of the objects is repeated. In general, the number of permutations of n objects taken r at a time is written nPr. Using the multiplication principle, $_nP_r = n(n-1)\ldots(n-r+1)$, which can be simplified to $\frac{n!}{(n-r)!}$. (See references [2] and [3].) On the TI graphics calculator, you can compute $_nP_r$ by first inserting n, pressing the $\boxed{\text{MATH}}$ button, moving to the PRB menu, and selecting **2:nPr**, and finally entering the value for r. For example, to find $_6P_3$ in the above problem, insert **6** on the text screen, locate **nPr** as described above, insert **3**, and press $\boxed{\text{ENTER}}$.

PROBLEM 6

Pierre, a customer at Blaise's Bistro, always has two desserts. How many possible selections can Pierre make if he is not concerned about the order in which he eats the desserts?

SOLUTION

Consider a simpler problem. Suppose for a moment that Pierre is concerned with the order. How many possible ordered arrangements are there?

Referring to the dessert menu, let K stand for Key Lime Pie, A for Apple Pie a la mode, B for Black Forest Cake, and F for French Vanilla Ice Cream. Then the possible ordered arrangements of four desserts taken two at a time is shown here:
KA, AK
KB, BK
KF, FK
AB, BA
AF, FA
BF, FB
Note that there are 12 arrangements, which could have been found using $_4P_2$.

Returning to Pierre's original problem, we observe that pairs of desserts are duplicated if order is not important. For example, KA and AK represent the same selection of two desserts if order is not relevant. Hence, the number of "paired selections" for Pierre is six. This is denoted by $_4C_2$ where C indicates combinations in which order does not matter.

COMMENTS

In general a selection of objects in which order is of no importance is called a *combination.* The number of combinations of n objects chosen r at a time is denoted by $_nC_r$ where $_nC_r = \frac{_nP_r}{r!} = \frac{n!}{[(n-r)!\,r!]}$. (See references [2] and [3].)

The above problem also indicates that combinations and permutations are related. In this case 6 (that is $_4C_2$) is equivalent to 12 (that is $_4P_2$) divided by 2 (that is 2!). In this example it is easy to see the duplicate pairs, and hence 2 is more obvious than 2!. However, if Pierre selected three desserts there would be 3! = 6 ways of ordering each selection of three desserts. For example, *KAB, KBA, ABK, AKB, BAK,* and *BKA* represent the set of all possible ordered arrangements of one selection of three desserts. For the case of three desserts, you would have $_4C_3 = \frac{_4P_3}{3!} = 4$.

On the TI graphics calculator, you can compute **nCr** by inserting n, pressing the $\boxed{\text{MATH}}$ button, moving to the PRB menu, selecting **3:nCr**, and finally entering the value for r. For example, in finding $_4C_2$ in the above problem, insert **4** on the text screen, locate **nCr** as described above, insert **2**, and press $\boxed{\text{ENTER}}$.

PROBLEM 7

Kelly orders from the A-La-Carte menu at Blaise's Bistro. How many selections are possible if Kelly chooses two green vegetables and three potato toppings to go with one selection of soup, meat, and dessert?

SOLUTION

In this problem it is assumed that order is of no interest to Kelly. For example, Kelly is not concerned about the order in which the green vegetables are placed on the plate, nor the order in which toppings are placed on the potato. Hence, the soup selection can be made in $_2C_1$ ways, as can the meat selection. For the green vegetables, two are to be selected from four. This can be done in $_4C_2$ ways. In a similar way, three potato toppings can be selected in $_5C_3$ ways, and the one dessert item can be selected in $_4C_1$ ways.

Because each selection is independent of the others, the multiplication principle applies, and the total number of selections for Kelly is $_2C_1 * _2C_1 * _4C_2 * _5C_3 * _4C_1 = 960$.

Using Counting Techniques in Probability

PROBLEM 8

In the Illinois lottery players choose six different numbers from the numbers 1 through 54. One set of six numbers is selected at random to determine the winning numbers. If a player chooses one set of six different numbers, what is the player's probability of selecting all six winning numbers?

SOLUTION

Because there are 54 numbers, there are $_{54}C_6$ ways of choosing the winning set of six numbers. Combinations are used rather than permutations because order is of no importance in the random selection of the winning numbers. For example the sets {1, 6, 18, 32, 37, 41} and {18, 6, 1, 41, 37, 32} represent the same selection.

The player has only one set of six numbers. Hence, the probability of the player winning is $\dfrac{1}{_{54}C_6}$ which is approximately 1 in 26 million.

CLASSROOM ACTIVITIES

The following activities may be useful in helping students develop a number of the counting principles considered in Module 3.

Activity 1—Another Birthday Problem

Pose the following questions: How many people are needed to insure that *at least two people* have the same birth month? to insure that *at least three people* have the same birth month?

Have students dramatize the situation by coming up to the front of the room one by one and announcing their birthday month. Determine how many were needed before two people had the same birthday month. If another group did this, would the number be the same? What is the worst possible scenario? This question can be answered by having the students dramatize a situation in which no birthday month is repeated until all months have been exhausted.

Activity 2—Arranging Students for a Photograph

Pose the following question: Suppose four people are to sit on four chairs placed in a line for a photograph. How many ways are there of arranging the four people?

Have four students simulate this situation, explaining their strategy as they go. Then have other groups suggest alternative strategies.

Activity 3—Arranging Students for Another Photograph

Pose the following question: Suppose there are five people available for a photograph, but only three of them will be chosen to sit on three chairs placed in a line. How many ways are there of selecting the three people to sit in the three chairs?

Have five students simulate this situation, explaining their strategy as they go. Then have other groups suggest alternative strategies.

Students are likely to adopt various strategies in solving this problem. Two common strategies are (1) to use a permutation approach, deciding how many ways each of the three chairs can be filled, or (2) to use a combination approach, selecting the three people and then order them on the seats. Decide how many ways this can be done.

IMPLEMENTATION GUIDELINES

Teaching Suggestions by Activity

Starting 3-1 Naming the Dog This activity sets the stage for future work. In preparation for the activity, it may be appropriate to review conventions governing "order of operations." Students may find using the calculator helpful in the *Reflect* section.

Starting 3-2 How Many Faces? It may be useful to set up a sample space of ordered pairs that enables all possible sums to be found. Discuss the sample space, highlighting some of the patterns that appear and noting, in particular, that outcomes such as (1,6) and (6,1) are different.

Starting 3-3 Birth Month in a Box Assuming a normal class size, the actual number of students in the class is irrelevant to this activity. After each draw, the question students should consider is: "What is the worst possible outcome?" The most general answer to the question posed in the activity is 13. If the number of months represented by the students in your class is less than 12, the answer will change. It will always be one more than the number of months represented. (See *Content Background* on page 46.)

Starting 3-4 Three the Same The general case for classes with birthdays in all 12 months is $12(n - 1) + 1$, where n is the number of names with birth month alike. The solution for your class depends on how many of the 12 months are represented. If any are missing, the 12 in the formula above changes to the number of months represented. Students may find the calculator useful for computation in this activity.

Starting 3-5 First Names in a Box Be sure to mix the names each time before they are drawn. After each draw the question students should consider is: "What is the worst possible outcome?"

Starting 3-6 How Many Sandwiches? A tree diagram or some other appropriate form of listing may be helpful for finding the solution to the question posed for this activity. You may find that the tree diagram helps students make the transition to the product rule for determining all possibilities. (See *Content Background* on page 44.)

Starting 3-7 How Many Kinds of Pizza? Because the product rule is used, students may find the calculator useful for computing the number of different meals in this activity.

Developing 3-1 Batting Order To prepare for this activity review the meaning of *factorial*. (See *Content Background* on page 48.)

Developing 3-2 Pitcher Bats Last Note that the restriction the pitcher must bat last changes the result from 9! to 8! because there is then no choice for the pitcher's position in the batting order.

Developing 3-3 Raffle Tickets To prepare for this activity review and discuss *permutation* (see Appendix E, "Word Bank," in the Student Edition). Stress the fact that order matters.

Developing 3-4 Calculating the Number of Raffle Tickets Permutations are again used in this activity, and the standard notation is now introduced. Stress that this activity involves dealing with arrangements in which order is important.

Developing 3-5 "Special" Raffle Tickets This activity involves both the use of permutations and the product rule.

Developing 3-6 "Extra-Special" Raffle Tickets This activity involves recognizing two different sets of permutations, combining them appropriately, and using the product rule. Filling in the boxes on the sheet may help students articulate the different possible arrangements.

Developing 3-7 Mini-Lotto In preparation for this activity it is appropriate to review *combinations* (see Appendix E, "Word Bank," in the Student Edition and *Content Background* on page 50). Stress that order does not matter for combinations.

Developing 3-8 Calculating the Mini-Lotto Pairs
Combinations are used in this activity. Remind students that because order does not matter for combinations (for example, 2-1 and 1-2 are the same combination), the formula essentially "takes out duplicates."

Developing 3-9 Pick Three Applying the combination formula involves division by 3!, to "take out" the number of different ways the three digits can be arranged.

Developing 3-10 Pick Three Again Students apply the combination formula to solve the problem posed in this activity.

Note: The three Extending activities of this module can be used as assessment projects.

Extending 3-1 Many the Same This activity is an extension of the pigeonhole principle considered in Starting 3-3 and Starting 3-4.

Extending 3-2 Two in the Same Month It may be helpful to review complements as this notion may be helpful to students in solving the problem in this activity. (See Content *Background on page 22.*)

Extending 3-3 The State Lottery and Extending 3-4 Winning Hands These two activities involve probabilities. Students use the counting ideas in this section to find the number of ways the person can win (this is the numerator in the probability ratio) and the total number of possible outcomes (this is the denominator of the probability ratio).

MODULE 4:
Modeling and Predicting

OVERVIEW AND PURPOSE

In this module students will determine *lines (or curves) of best fit* to model sets of two-variable data that are inherently linear or non-linear. The main purpose for finding these best-fit models of real-world data is to make optimum predictions for the dependent variable (y) for various values of the independent variable (x). You can use the TI graphics calculator to visually place lines of best fit by inspecting the scatter plot and using the DRAW menu. The calculator also enables this process to be carried out analytically by determining *least-squares* equations. In this module students will learn how to:

- draw scatter plots for two-variable data;

- draw lines of best fit using a strictly visual approach facilitated by the STAT DRAW menu on the TI graphics calculator;

- use the TI graphics calculator to determine least-squares linear equations for various sets of two-variable data;

- interpret residual plots and indices such as the mean square error (MSE) and the linear correlation coefficient (r) and use them as criteria for evaluating goodness of fit for models applied to two-variable data;

- apply various transformations to data in order to achieve better models and to determine the equations of these best-fit models using both the transformed and the original values; and

- use best-fit models to make predictions for both linear and nonlinear data.

The background presented here is intended to assist you in planning instruction for this module. You may also incorporate some of the background ideas directly into the instructional program you use to introduce this topic.

OUTLINE OF KEY MATHEMATICAL IDEAS BY ACTIVITY

Activity Title	Key Mathematical Ideas
Starting	
4-1 Shoe and Arm Lengths	scatter plot, line of best fit
4-2 Shoe and Name Lengths	linear relationship
4-3 The Missing Movie	linear relationship
Developing	
4-1 Shoe Lengths and Consonants	line of best fit, error
4-2 The Line of Best Fit	line of best fit, mean square error
4-3 Equation for the Line of Best Fit	line of best fit, correlation
4-4 Graphing on the Scatter Plot	line of best fit, correlation
4-5 Finding the Correlation	correlation
4-6 Money Made and Movie Ranking	correlation
4-7 Money Made and Year Released	correlation
4-8 Forecasting the Missing Movie	predicting and forecasting
Extending	
4-1 Shoe Lengths and Arm Lengths	predicting
4-2 Testing the Radius and Area Relationship	nonlinear relationships
4-3 Transforming the Radius and Area Data	transforming nonlinear data
4-4 Fitting Radius to Area	nonlinear best fit
4-5 Testing the Radius and Volume Relationship	nonlinear best fit
4-6 Crude-Oil Production	testing linearity
4-7 Transforming the Crude-Oil Production Data	transforming nonlinear data
4-8 Fitting Year and Crude-Oil Production	nonlinear best fit
4-9 An Alternative Transformation	transforming nonlinear data
4-10 Another Transformation	transforming nonlinear data
4-11 Non–English-Speaking Americans	finding a relationship
4-12 How the United States Has Grown	finding a relationship

Calculator Helps

For Module 4:

Calculator Help 6: Using the Line Function; Cursor Movement in the Line Function

Calculator Help 7: Finding the y-Intercept, the Slope of the Line of Best Fit, and the Correlation Coefficient (r); Using the Equation to Draw the Line of Best Fit

Calculator Help 8: Using the Best-Fit Equation to Predict a y Value for a Given x

Calculator Help 9: Entering Transformed Data (TI-82 calculators only)

CONTENT BACKGROUND

The *Content Background* for Module 1 discussed the construction of scatter plots and the use of the linear correlation coefficient as an index of how well points on a scatter plot conform to a straight-line relationship. In this module students construct lines or curves of best fit to represent the two-variable data displayed in the scatter plots. The following first considers the case of fitting a straight line, as this is typically the first model used to represent two-variable data. This modeling process provides the background for considering nonlinear models. The problems are drawn from used car data gathered by the authors in January 1995 (Figure 1).

Figure 1

Year of Production and Price for Used Honda Accords

x: Age of Car (Model Year)	y: Price (in Dollars)
5 (1990)	$11995
5 (1990)	9900
6 (1989)	9000
6 (1989)	8000
7 (1988)	6800
7 (1988)	6000
7 (1988)	8700
7 (1988)	5650
7 (1988)	7300
7 (1988)	7000
8 (1987)	6999
9 (1986)	5500
10 (1985)	3500
15 (1980)	1000
16 (1979)	1500

Determining Linear Models

PROBLEM 1

Determine a line of best fit for the Honda Accord used car prices shown in Figure 1 where the independent variable, x, is the age of the car, and the dependent variable, y, is the used car's price. Use your line to estimate the price of a 1984 Honda Accord.

SOLUTION

Four solution methods are considered for this problem. The first method is intuitive—it is based strictly on visual estimation. Essentially, it locates the line that is closest to the data points. The process does, however, foreshadow the next two solution methods that are based on minimizing the *error* between the actual data points and the line of best fit.

More specifically, for each data point (x,y) on the scatter plot, the *error value*, or *residual*, is the difference between each given y value and the y value estimated by the line of best fit for the corresponding x value. That is, the error for each point on the scatter plot is $y - y'$, where y' is the estimated y value. This is illustrated in Figure 2. (See references [1], [5], [7], [8], [11], [12].)

Figure 2

Graphical
Representations of
Residuals

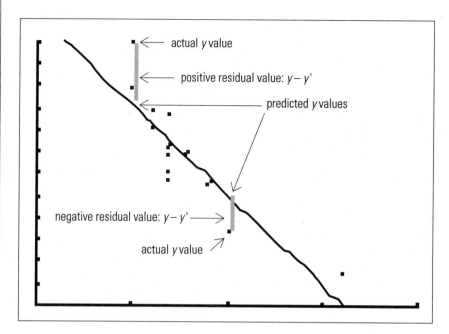

In practice, because $y - y'$ can be positive or negative, the *sum of squares of the errors* (SSE) is used as a criterion for determining the line of best fit. (Some authors prefer to use the *mean square error*. See references [5] and [11].) Using the SSE criterion and methods of calculus, the slope and intercept of the line of best fit can be determined. This line of best fit is called the *least-squares regression line*. It passes through the *centroid* of the data set, which is the point on the graph whose coordinates are the means of x and y: $(\bar{x}, \bar{y})$. This characteristic of the least-squares regression line is very useful when locating lines of best fit with visual estimation and is illustrated in Method 2 that follows. After conceptualizing this process, we will use the TI graphics calculator and its built-in programs for determining the slope and intercept of the least-squares regression line.

PROBLEM 1, METHOD 1: VISUAL ESTIMATION WITH THE TI GRAPHICS CALCULATOR

The basis for this method is that a reasonable best-fit model can be found by drawing a line that is "closest to most of the points." Using the calculator you can locate such a line. Press [2nd] [DRAW] and then use the arrow keys to select **2:Line(**. Follow these steps to draw a line:

1. Create a scatter plot of the data (see "Calculator Help 4: Generating a Scatter Plot" in Appendix B of the Student Edition).

2. Press [2nd] [DRAW].

3. Select **2:Line(**. You will see a blinking cursor on the TI-81 calculator.
4. Use the arrow keys to locate the cursor where you want the left endpoint of your line of best fit. Press [ENTER].
5. Use the arrow keys to extend the line until it is located in your best-fit position. Press [ENTER] to fix the right endpoint. (For additional information on this sequence of calculator keystrokes, see "Calculator Help 6: Using the Line Function; Cursor Movement in the Line Function" in Appendix B of the Student Edition.)

Figure 3a
Regression Line Drawn
Using TI Graphics
Calculator DRAW Menu

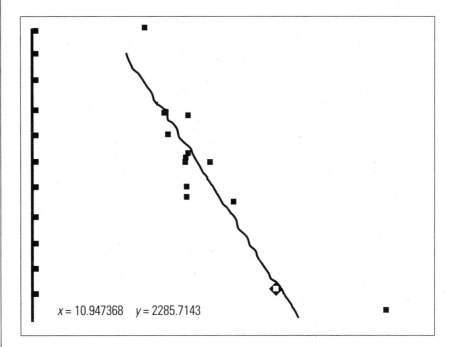

$x = 10.947368$ $y = 2285.7143$

After positioning the line, you can use it to estimate the price of a 1984 Honda Accord, a car 11 years old. Use the arrow keys to move the cursor until the x-coordinate, shown in the bottom of the screen, is approximately 11. Now move the cursor vertically to the line of best fit and read the x-coordinate. In the graphics screen shown in Figure 3a, the estimated price for a 1984 Accord is $2286.

This is a useful intuitive process, particularly when working with middle-school students. Students should be encouraged to identify the criteria used for establishing a line of best fit. This may lead naturally to a discussion of error or residuals.

PROBLEM 1, METHOD 2: VISUAL ESTIMATION THROUGH THE CENTROID OF THE DATA

In this solution method visual estimation is used to locate several estimates of the line of best fit. All these lines pass through the centroid. The equation for each of these lines is found by first locating graphically a second point on a visually estimated line.

Next, the slope of this line is found from the coordinates of the centroid and the second point. Recall that the formula for slope is $\frac{(y_2 - y_1)}{(x_2 - x_1)}$ Once the slope is determined, use the form $y = mx + b$ to find the equation.

For example, for a visually estimated line that passes through the centroid (8.13, 6596.27) and the x-axis at the point (16,0), the slope is –838.15. (See figure 3b. Note the choice of an axis point to go with the centroid.) Hence, the equation is of the form $y = -838.15x + b$. Substituting (16,0), because this point is on the line, we have $0 = -838.15(16) + b$. Thus, the value of b is 13,410.4. Finally, the equation is $y' = -838.15x + 13,410.4$, where ordered pairs (x, y') represent points on the visually estimated best-fit line.

Figure 3b

Visually Estimated Line of Best Fit

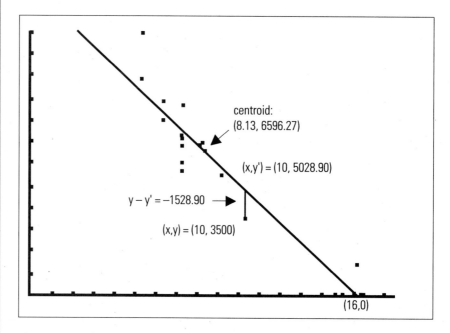

centroid:
(8.13, 6596.27)

$(x, y') = (10, 5028.90)$

$y - y' = -1528.90$

$(x, y) = (10, 3500)$

(16,0)

The sum of squares of the errors (SSE) is then calculated for each line using the formula $\Sigma(y - y')^2$ where y is the original data value for a given x and y' is the predicted or best-fit line value for that same x. For example, for the point (10,3500) on the scatter plot, $y' = -838.15(10) + 13410.4 = 5028.90$. Hence, the residual $y - y'$ for this point on the scatter plot is $3500 - 5028.90 = -1528.9$ (see figure 3b). Its contribution to the SSE is $(-1528.9)^2 = 2,337,535.21$. In a similar way, the residuals for each of the other points on the scatter plot are found and the squares of the residuals are added to get the SSE for this visually estimated best-fit line. In this example, the SSE is 21,601,193.36.

The SSE for this visually estimated best-fit line is compared with the SSE for other visually estimated lines. The equation with the least SSE is considered to be the line of best fit under this method. Figure 4 presents the slopes and SSE for several visually estimated lines that pass through the centroid of the used car data.

Figure 4
Honda Used Car Data: Linear Models Sum of Squared Errors for Various Slopes

Slope	Sum of Squared Errors (SSE)
−870	21,916,770
−860	21,784,935
−850	21,683,040
−840	21,611,100
−830	21,569,100
−820	**21,557,055**
−810	21,574,950
−800	21,622,800
−790	21,700,590
−780	21,808,320
−770	21,946,005

In this case, the best approximation to the least-squares regression line is the line with slope −820 because it has the least SSE. Using the fact that this passes through the centroid (8.13, 6596.27), the equation of this line is $y' = -820x + 13265.6$.

In building an understanding of the least-squares process using this method, it is useful to have students work in small groups. Each group can calculate the centroid, draw a scatter plot that includes the centroid, visually locate a line of best fit through the centroid, determine its equation from the scatter plot, and calculate the SSE. A table of values, similar to the one above, can be constructed using each group's data. By noting the least SSE in the table, students can identify the corresponding slope of a least-squares regression line. Among the lines generated by the student groups, this is the best approximation for the least-squares line. (See reference [11].)

PROBLEM 1, METHOD 3: THE LEAST-SQUARES METHOD
The graphics calculators have facility for finding the equation of the least-squares regression line and the corresponding correlation coefficient. While the TI-81 graphics calculator allows for predicted y-values to be found for given values of x, it does not provide an efficient way to calculate the residuals or the SSE. On the other hand, the TI-82 graphics calculator has excellent facility for calculating residuals and the SSE.

For help in using the TI graphics calculators to determine the regression equations and in plotting them on scatter plots of data, see "Calculator Help 7: Finding the y-Intercept, the Slope of the Line of Best Fit, and the Correlation Coefficient (r); Using the Equation to Draw the Line of Best Fit" in Appendix B of the Student Edition.

Figure 5 shows the results of the calculations as viewed on the TI-82 graphics calculator. (Note that the variables a and b are used in reverse on the TI-81 graphics calculator; that is, a represents the y-intercept and b represents the slope.)

Figure 5

TI-82 Screen: Results of
Linear Regression
Calculation

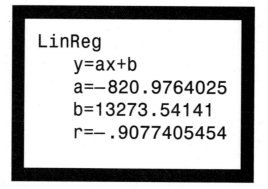

Hence the equation of the least-squares regression line is $y' = -820.98x + 13273.54$. Using this line the best-fit prediction for the price of a 1984 Honda Accord (11 years old) is \$4242.80. The line of best fit is shown in Figure 6.

Figure 6

Honda Accord Used Car
Data: Scatter Plot and Plot
of Least-Squares Linear
Regression Line

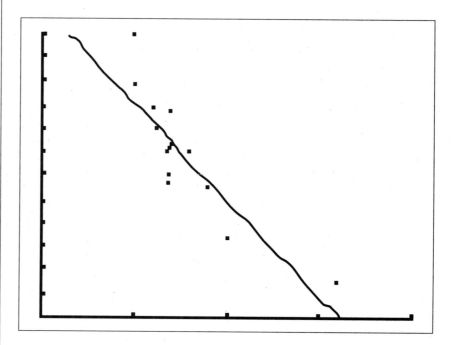

The question naturally arises: How good is this prediction? This, in turn, depends on how well the linear model fits the data. Four criteria can be used to evaluate *goodness of fit*.

PROBLEM 1, CRITERIA FOR EVALUATING GOODNESS OF FIT

(1) Scatter Plot. As discussed in the Module 1 *Content Background* on pages 8–9, the shape, strength, and direction of a scatter plot provide visual indicators of the goodness of fit of a linear model. In these activities students are expected to rely heavily on visual analysis of scatter plots to evaluate goodness of fit. The scatter plot for the used car data suggests a decreasing relationship that is more curvilinear than linear. This is illustrated by the curve on the scatter plot in Figure 7.

Figure 7

Honda Accord Used Car Data: Superimposed Plot Suggesting Curvilinear Nature of the Scatter Plot

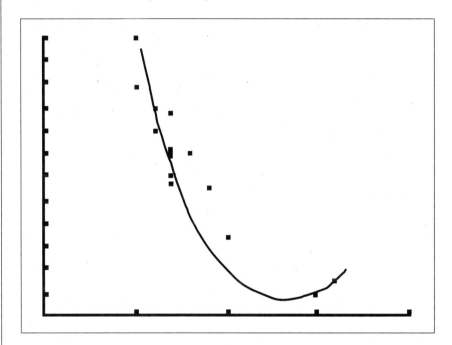

(2) Correlation Coefficient. (See Figure 9, page 10, in Module 1.) The correlation coefficient is readily available when using the TI graphics calculators. It is an efficient, but not always sensitive, measure of goodness of fit. When the correlation coefficient is close to −1 or +1, there is evidence of a strong linear relationship. For the used car data, the correlation coefficient of −0.908 indicates a strong, but not perfect, linear fit.

(3) Sum of Squares of the Errors (SSE). This is a precise method for evaluating the goodness of fit. It is, however, extremely arduous, even when using the TI-82 graphics calculator. The SSE for the used car data is determined from information shown in Figure 8. We have used the list feature of the TI-82 graphics calculator to determine y' (the predicted values of y), the residual or error values $(y - y')$, and finally the square of each error $[(y - y')^2]$.

L_1 $x:$ age	L_2 $y:$ price	L_3 y'	L_4 $y - y'$	L_5 $(y - y')^2$
5	$11995	9168.66	2826.34	7,988,201
5	9900	9168.66	731.34	534,859
6	9000	8347.68	652.32	425,517
6	8000	8347.68	−347.68	120,883
7	6900	7526.71	−626.71	392,761
7	6000	7526.71	−1526.71	2,330,833
7	8700	7526.71	1173.29	1,376,617
7	5650	7526.71	−1876.71	3,522,028
7	7300	7526.71	−226.71	51,396
7	7000	7526.71	−526.71	277,420
8	6999	6705.73	293.27	86,007
9	5500	5884.75	−384.75	148,035
10	3500	5063.78	−1563.78	2,445,400
15	1000	958.90	41.10	1,690
16	1500	137.92	1382.08	1,855,265

sum of squared errors: 21,556,913

As discussed previously, although the correlation coefficient for the used car data is close to 1 and thus suggests a strong linear relationship, the SSE indicates that there is still significant residual error associated with the linear model.

(4) Residual Plot. The residual plot is constructed by plotting the residual error $(y - y')$ against its corresponding x value. In this way residuals remain positive or negative. The residual plot, based on the least-squares regression line, is shown in Figure 9 for the used car data. Figure 8 provides the residual errors $(y - y')$ and the x values to create the residual plot. This is carried out by using the TI-82 graphics calculator scatter plot option in [2nd] [STAT PLOT] and setting the x list to L_1 and the y list to L_4, assuming the appropriate data are in those lists.

When using a residual plot to evaluate goodness of fit, a strong fit is indicated by an *absence* of pattern, that is, a random scattering of points. For the residual plot in Figure 9, observe that for newer Honda Accords (smaller values of x), the residuals tend to be positive. This is also true for the oldest Honda Accords. Cars in the middle age range tend to have negative residuals. While this is indicative of some existing pattern, it also should be noted that there are residuals that deviate from this apparent pattern.

Figure 9
Honda Accord Used Car
Data: Residual Plot

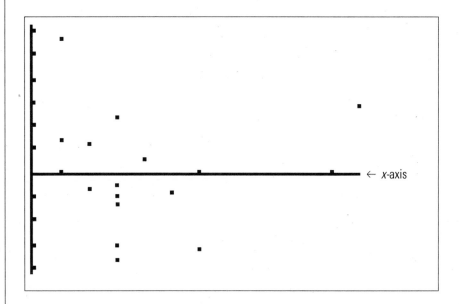

Therefore, the residual plot tends to confirm the earlier analysis that the linear model, while satisfactory, may not be the best model for the used car data. After considering the median-median regression line, you will explore a number of nonlinear models to determine whether they produce a more effective fit to the used car data. (For further background on criteria for evaluating goodness of fit, see references [1], [5], [8], [11], and [12].)

PROBLEM 1, METHOD 4: MEDIAN-MEDIAN LINE OF BEST FIT
This method is not considered in the Student Edition but is extensively used in many texts. The method essentially fits a line to representative median-median points of the data. The TI-81 graphics calculator does not provide facility for determining the median-median line of best fit. However, the TI-82 graphics calculator does include a median-median model with the other regression equations. For a thorough discussion of the median-median method, see references [1], [4] and [7].

Using the TI-82 graphics calculator, the median-median line of best fit for the used car data is $y' = -1322.22x + 16550$. Based on this model the best-fit prediction for the price of a 1984 Honda Accord is $2005.56. The SSE for the median-median line is 68,785,198, compared to an SSE of 21,556,913 for the least-squares regression line. This result exemplifies the key property of the least-squares regression line. That is, of all possible regression lines, the least-squares regression line has the minimum SSE.

Determining Nonlinear Models

As discussed in the previous section, the least-squares linear regression line produced a satisfactory model for the used car data. However, the evaluation of the goodness of fit indicates that there may be a better model, perhaps one that is nonlinear.

Several nonlinear models can be investigated, including exponential, power (quadratic), and logarithmic. The *Content Background* for Module 1 on page 11 includes scatter plots to illustrate some of these models, and this is illustrated again in Figure 10. Problems 2, 3, and 4 investigate nonlinear models for the Honda Accord used car data.

Figure 10

Examples of Nonlinear Models

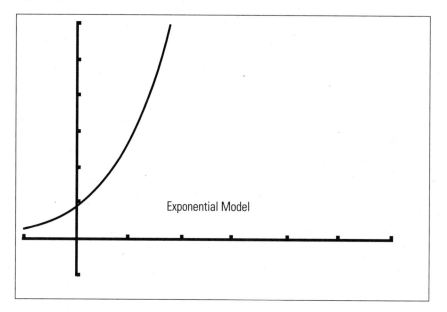

Exponential Model

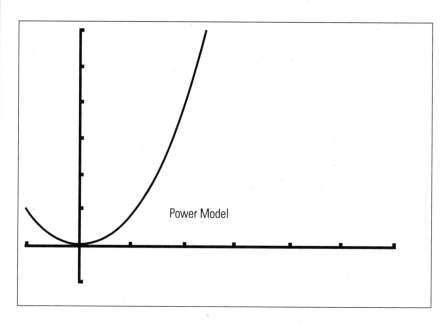

Power Model

Figure 10 (cont.)

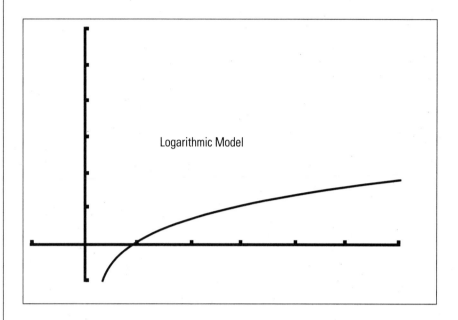

Logarithmic Model

PROBLEM 2

Use a natural log transformation on the data to find a best-fit model for the used car data.

SOLUTION

This transformation is normally used when a scatter plot suggests that a best-fit model for the data would be of the form $y = a \ln x + b$. For this reason, it is typically called the *logarithmic* model. Observation of the Honda Accord used car scatter plot suggests the data are not best modeled by a logarithmic pattern, although we must remember that the coefficient a may be less than 0. To carry out the transformation let $x_1 = \ln x$ and determine the values of x_1 corresponding to x, as shown in Figure 11.

Figure 11
Honda Accord Used Car
Data: Logarithmic Model
Data Values

x: age	y: price	x_1: $\ln x$
5	$11995	1.6094
5	9900	1.6094
6	9000	1.7918
6	8000	1.7918
7	6900	1.9459
7	6000	1.9459
7	8700	1.9459
7	5650	1.9459
7	7300	1.9459
7	7000	1.9459
8	6999	2.0794
9	5500	2.1972
10	3500	2.3026
15	1000	2.7081
16	1500	2.7726

Sum of squared errors: 13,311,677

Use the calculator and the ordered pairs (x_1, y) to determine the least-squares linear model relating x_1 and y. The TI-82 graphics calculator screen in Figure 12 shows the slope (a), the y-intercept (b), and the correlation coefficient (r) for this model.

Figure 12

TI-82 Screen: Results of Regression Calculation for Logarithmic Transformation

```
LinReg
   y=ax+b
   b=23282.26426
   a=-8196.085119
   r=-.9440939048
```

Therefore, the equation of the best-fit linear model for the transformed data (x_1, y) is $y' = -8196.09x_1 + 23{,}282.26$. The best-fit model for (x,y), the original data, is found by using the substitution $x_1 = \ln x$.

The resulting equation is $y' = -8196.09 \ln x + 23{,}282.26$. Because you use the original transformation $x_1 = \ln x$ in reverse, the best-fit model for the original (x,y) data is often called the *retransformed* equation. As you might expect, the retransformed equation is a natural log function.

Based on this model, the best-fit prediction for the price of a 1984 Honda Accord is \$3628.91. Note that the correlation coefficient is -0.944 and that the SSE for this transformation is 13,311,677. In terms of the four goodness-of-fit criteria, observe that (1) the scatter plot does not appear to follow a logarithmic pattern; (2) the correlation coefficient for the transformed data is stronger than for the linear model; (3) the SSE is significantly less than the SSE for the linear model; and (4) the residual plot, shown in Figure 13, is reasonably random. That is, the residual plot does not follow a systematic pattern. In summary, the log transformation provides a better model for the used car data than the linear model.

Figure 13

Honda Accord Used
Car Data: Logarithmic
Residual Plot

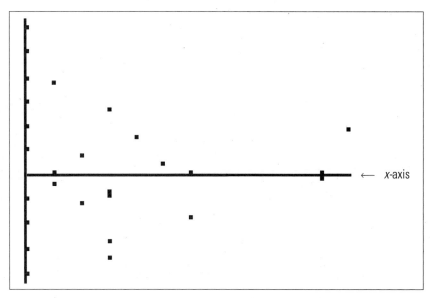

$\leftarrow$ *x*-axis

PROBLEM 3

Use an exponential transformation on the *y* data to find a best-fit model for the used car data.

SOLUTION

This transformation is normally used when a scatter plot suggests that a best-fit model for the data is exponential, that is, of the form $y = ae^x$. For this reason, it is typically called the *exponential* model. Observation of the Honda Accord used car scatter plot suggests that the data do follow an exponential pattern.

To carry out the transformation, let $y_1 = \ln y$ and determine the values of y_1 corresponding to *y*, as shown in Figure 14.

Figure 14

Honda Accord Used Car
Data: Exponential Model
Data Values

x: age	*y*: price	y_1: ln*y*
5	$11995	9.3922
5	9900	9.2003
6	9000	9.1050
6	8000	8.8972
7	6900	8.8393
7	6000	8.6995
7	8700	9.0711
7	5650	8.6394
7	7300	8.8956
7	7000	8.8537
8	6999	8.8535
9	5500	8.6125
10	3500	8.1605
15	1000	6.9078
16	1500	7.3132

Sum of squared errors: 11,385,123

Use the calculator and the ordered pairs (x, y_1) to determine the least-squares linear model relating x and y_1. The TI-82 screen in Figure 15 shows the slope (a), the y-intercept (b), and the correlation coefficient (r) for this model.

```
LinReg
   y=ax+b
   a=-0.203328
   b=10.28873
   r=-.9687066141
```

Therefore, the equation of the best-fit linear model for the transformed data (x, y_1) is $y_1 = -0.20328x + 10.28873$. The best-fit model for (x, y), the original data, is found by using the substitution $y_1 = \ln y$.

$$y_1 = -0.20328x + 10.28873$$
$$\ln y = -0.20328x + 10.28873$$
$$y = e^{-0.20328x + 10.28873}$$
$$y = e^{-0.20328x} * e^{10.28873}$$
$$y = 29399.41338e^{-0.20328x}$$

Based on the retransformed equation, the best-fit prediction for the price of a 1984 Honda Accord is $3142.12. Note that the correlation coefficient is -0.969 and that the SSE for this transformation is 11,385,123. In terms of the four goodness-of-fit criteria, observe that (1) the scatter plot does appear to follow an exponential pattern; (2) the correlation coefficient for the exponential model is stronger than the one for either the linear model or the logarithmic model; (3) the SSE is significantly less than the SSE for the linear model and is somewhat less than the SSE for the logarithmic model; and (4) the residual plot, shown in Figure 16, is quite random. In summary, the exponential model is a better model for the used car data than either of the previous two regression models.

Figure 16
Honda Accord Used
Car Data: Exponential
Residual Plot

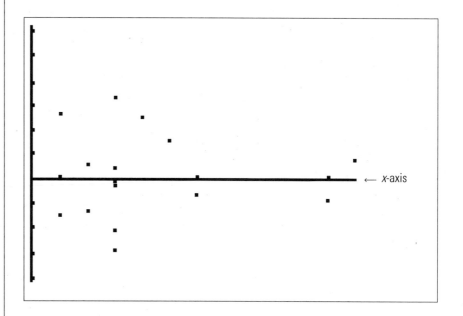

PROBLEM 4

Use natural log transformations on both the x and y data to find a best-fit model for the used car data.

SOLUTION

This transformation is normally used when a scatter plot suggests that a best-fit model for the data is a power function, that is, of the form $y = ax^b$. For this reason it is typically called the power model. Observation of the Honda Accord used car scatter plot suggests that the data may follow the pattern of a power function.

To carry out the transformation, let $x_1 = \ln x$ and let $y_1 = \ln y$. Determine the values of x_1 and y_1 corresponding to x and y, respectively as shown in Figure 17.

Figure 17
Honda Accord Used
Car Data: Power Model
Data Values

x: age	y: price	x_1: ln$x$	y_1: ln$y$
5	$11995	1.6094	9.3922
5	9900	1.6094	9.2003
6	9000	1.7918	9.1050
6	8000	1.7918	8.8972
7	6900	1.9459	8.8393
7	6000	1.9459	8.6995
7	8700	1.9459	9.0711
7	5650	1.9459	8.6394
7	7300	1.9459	8.8956
7	7000	1.9459	8.8537
8	6999	2.0794	8.8535
9	5500	2.1972	8.6125
10	3500	2.3026	8.1605
15	1000	2.7081	6.9078
16	1500	2.7726	7.3132

Sum of squared errors: 21,597,056

Use the calculator and the ordered pairs (x_1, y_1) to determine the least-squares linear model relating x_1 and y_1. The TI-82 screen in Figure 18 shows the slope (a), the y-intercept (b), and the correlation coefficient (r) for this model.

```
LinReg
    y=ax+b
    a=-1.924251991
    b=12.55288
    r=-.9552966897
```

Therefore, the equation of the best-fit linear model for the transformed data (x_1, y_1) is $y_1 = -1.92425x_1 + 12.55288$. The best-fit model for (x,y), the original data, is found by using the substitutions $x_1 = \ln x$ and $y_1 = \ln y$.

$$y_1 = -1.92425x_1 + 12.55288$$
$$\ln y = -1.92425\ln x + 12.55288$$
$$\ln y = \ln x^{-1.92425} + \ln 282908.8387$$
$$\ln y = \ln(x^{-1.92425} * 282908.8387)$$
$$y = 282908.8387 * x^{-1.92425}$$

Based on the retransformed model, the best-fit prediction for the price of a 1984 Honda Accord is \$2803.77. Note that the correlation coefficient is -0.955 and that the SSE for this transformation is 21,597,056. In terms of the four goodness-of-fit criteria, observe that (1) the scatter plot appears to follow the pattern of a power function; (2) the correlation coefficient for the power model is stronger than the one for either the linear model or the logarithmic model, but slightly less than for the exponential model; (3) the SSE is greater than the SSE for either the logarithmic or exponential models and about the same as for the linear model; and (4) the residual plot, shown in Figure 19, does not appear to be as random as the residual plots for the other nonlinear models. In summary, the power model is certainly not as good a model of the Honda Accord used car data as the exponential model, and is not as good as the logarithmic model.

Figure 19
Honda Accord
Used Car Data:Power
Residual Plot

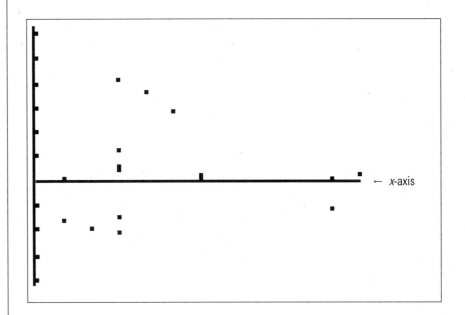

The conclusions about goodness-of-fit, particularly the superiority of the exponential model, appear to be confirmed by the scatter plots shown in Figure 20. These plots of the original data show the retransformed models for the linear and the three nonlinear models.

Figure 20
Honda Accord Used Car
Data: Scatter Plots with
Various Regression
Models Shown

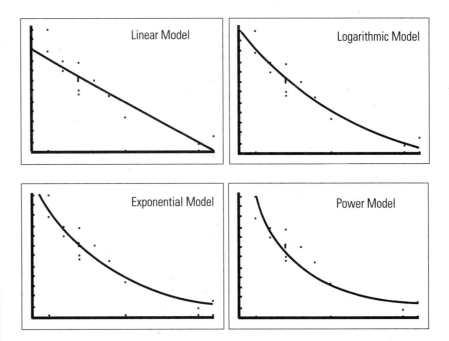

Summary

1. The use of the four goodness-of-fit criteria normally allows one to determine a best-fit model. However, in some instances the four criteria give conflicting messages, as tended to occur in the power model derived above. One Student Edition activity, Extending 4-12, is noteworthy in that no model is completely satisfactory in meeting all four goodness-of-fit criteria.

2. The discussion above considers some of the most frequently applied transformations. However, there is an entire family of transformations associated with polynomial regression models. The TI-82 has the capacity to determine quadratic, cubic, and quartic regression models. Students may wish to apply polynomial models in the case of Extending 4-12.

3. In the solutions to the problems presented above, this process was followed:

   - Use transformations $x_1 = \ln x$ and $y_1 = \ln y$ to transform (x,y) data to (x_1,y), (x,y_1), or (x_1,y_1) data.

   - Determine the linear regression equation for the transformed data, for example, (x_1,y).

   - Find the (x,y) or retransformed model by using the appropriate inverse transformation on the linear regression equation.

   This process enables students to develop a conceptual understanding of the use of transformations. Note, however, that the TI-81 and the TI-82 have the facility to determine logarithmic, exponential, and power models directly. For the TI-81 these models are under the CALC menu, accessed by pressing [2nd] [STAT]. With the TI-82, these models are under the CALC menu, selections 0, A, and B, accessed by pressing [STAT].

4. In the Student Edition the concept of *mean square error* (MSE) is introduced and used in the activities rather than sum of the squared errors (SSE). The MSE is, of course, related to the SSE and in the maximum likelihood sense, $\text{MSE} = \frac{\text{SSE}}{n}$. The Student Edition uses this maximum likelihood formula because it is much easier for students to conceptualize the notion of "mean" square error. However, more advanced textbooks use the unbiased MSE, $\text{MSE} = \frac{\text{SSE}}{(n-2)}$. See references [5] and [11] for details on the MSE.

5. For further background on fitting linear and nonlinear models, see references [1], [5], [7], [8], [11], and [12].

CLASSROOM ACTIVITIES

The following activities may be useful in helping students to develop some ideas on the relationships that exist in two-variable data.

Activity 1—Keeping Track of the Sum

One student rolls a die for the class. Keep a running total of the numbers rolled and record them in a table such as the following:

Roll Number	Running Total
1	4
2	9
3	13

Pose the question: What relationship, if any, exists between "Roll Number" (x) and "Running Total" (y)? Would the relationship change for different samples of data?

Activity 2—Population Growth in the United States

Show an overhead transparency table of the population in the United States from 1790 to 1994 (See "Data Set 13: Population of the United States from 1790–1990" in Appendix D of the Student Edition). Pose the following questions: Would you expect the data to be linear? Why or why not? What kind of relationship best describes the data? Explain.

Activity 3—The Bungee Jump

Present the following problem: Consider the flight of a bungee jumper from the moment the person leaves the platform until the lowest point is reached. If you collected data on the height of the bungee jumper above the ground at various times through this flight, would you expect the relationship between height and time to be linear? Why or why not? What kind of relationship best describes the data? Explain.

Note: Students could actually collect data by using a stuffed animal attached to an elastic cord to simulate the bungee jump.

IMPLEMENTATION GUIDELINES

Teaching Suggestions by Activity

Starting 4-1 Shoe and Arm Lengths Arm length can be defined as the distance from the elbow to the tip of the middle finger. The points in the graph are likely to be fairly clustered and approximate a straight line. When estimating the line of best fit, students should account for the endpoints of the data, not just those that are clustered in the center of the data.

Starting 4-2 Shoe and Name Lengths The points in this activity are likely to be broadly spread and will not generally approximate a straight line. (See *Content Background* for Module 1 on page 9.)

Starting 4-3 The Missing Movie One of the major reasons for modeling data with "best-fit" lines or curves is to make predictions. This idea can be emphasized through the *Reflect* activity.

Developing 4-1 Shoe Lengths and Consonants It may be helpful in this activity if you encourage students to compare their approaches with those of other students.

Developing 4-2 The Line of Best Fit Discuss mean square error and the fact that the error (vertical distance from the line of best fit) is squared, otherwise the positive and negative values would sum to 0. Students can determine the slope and intercept of the line of best fit either graphically or algebraically. Students may find it useful to create a table of values with these column headings: y, y', $y - y'$, and $(y - y')^2$. It may be helpful for students to make transparencies showing their equations and error tables as a basis of comparison and discussion. You might prefer to use the formula $\frac{(y - y')^2}{(n - 2)}$, but it is difficult to justify its use to students.

Developing 4-3 Equation for the Line of Best Fit Students use a calculator in this activity to find the slope and intercept of the line of best fit. The calculator performs calculations to find the line with the least mean square error.

Developing 4-4 Graphing on the Scatter Plot Students should find the regression equation first, then draw the scatter plot and superimpose the line of best fit on top of it. (See "Calculator Help 7: Finding the y-Intercept, the Slope of the Line of Best Fit, and the Correlation Coefficient (r); Using the Equation to Draw the Line of Best Fit" in Appendix B of the Student Edition.)

Developing 4-5 Finding the Correlation To prepare for this activity discuss the *correlation coefficient* (*r*). (See Appendix E, "Word Bank," in the Student Edition and *Content Background* for Modules 1 and 4 in the Teacher's Guide, pages 9 and 63.) Students obtain the value of *r* from the CALC menu.

Developing 4-6 Money Made and Movie Ranking Students should be sure to check the range whenever they enter a new set of data. You might also consider linear functions and note their relationship to the ideas of correlation. A true linear function would always have a correlation of 1 or −1.

Developing 4-7 Money Made and Year Released Students using TI-82 graphics calculators can put the "year" data in L_3. They would then need to make appropriate changes to <SetUp> in the CALC menu and also in the [STAT PLOT] conditions. Note that the two relationships, "movie ranking—money made" and "year released—money made" have negative correlations, the second being much weaker.

Developing 4-8 Forecasting the Missing Movie This activity involves using the best-fit equation to predict. Some of the factors that would affect the accuracy of this prediction are the recency of this year's top movie and inflation.

Extending 4-1 Shoe Lengths and Arm Lengths To indicate the goodness of their predictions, students should examine the scatter plot and the correlation coefficient (*r*). Be sure students understand that while predictions can be made beyond the range of the data, the line of best fit is a good predictor ONLY within the range.

Extending 4-2 Testing the Radius and Area Relationship Students should examine the correlation coefficient (*r*) to determine the goodness of their predictions. They should be aware that they are comparing a linear measure, *r*, with an area measure involving r^2. They might expect there will not be a very good linear fit.

Extending 4-3 Transforming the Radius and Area Data In this activity the data are transformed so they can be fitted by a straight line. Note that the transformation $y_1 = \sqrt{y}$ produces a quadratic regression model.

Extending 4-4 Fitting Radius to Area Students should get approximately the same results from the original equation and the transformed equation of the *Organize* section.

Extending 4-5 Testing the Radius and Volume Relationship
Providing that the data collected on the measures of the spheres are reasonably accurate, the produced value for volume, when r is 1.5, should be approximately equal to the volume found from the formula.

Extending 4-6 Crude-Oil Production One possibility is to start by plotting residuals. The more random the plot of residuals, the better the regression model.

Extending 4-7 Transforming the Crude-Oil Production Data, Extending 4-8 Fitting Year and Crude-Oil Production, Extending 4-9 An Alternative Transformation, and Extending 4-10 Another Transformation It would be helpful to discuss or review exponential and logarithmic functions and their inverse relationships. Also consider power functions. (See *Content Background* on pages 66–74.)

Extending 4-11 Non–English-Speaking Americans and Extending 4-12 How the United States Has Grown These activities would be appropriate as assessment projects for students working in groups or independently. See the comment about Extending 4-12 in the *Content Background* on page 74.

Selected Answers

Module 1: Analyzing Data

Starting 1-1: What's in a Name?

Reflect

- Various answers possible, but consider range, clusters, gaps, and extremes for the number of consonants data.
- Group the data, for example, 1-2, 2-4, 3-6, where 1-2 means that 1-consonant names have a frequency of 2.

Starting 1-2: Names in a Histogram

Reflect

- Frequency bars would be higher, but shape would stay much the same.
- Yes.

Starting 1-3: What's the Mean for the "Name" Data?

Organize

- Mean is 3.68 when using "Data Set 1: First Names of Students."

Reflect

- Extreme values influence the mean.

Starting 1-4: What's the Median for the "Name" Data?

Organize

- Median is 4.0.

Communicate

- For the "Even" data set—median is the mean of two middle scores. For the "Odd" data set—median is the middle score.

Reflect

- The median is generally not affected by extreme scores. The reordering may have a marginal effect.

Starting 1-5: Quartering the "Name" Data

Organize/Communicate

- Lower quartile (Q_1) = 3; upper quartile (Q_3) = 4.5.

Reflect

- Quartiles are generally not affected by extreme scores. The reordering may have a marginal effect.

Starting 1-6: What's the Mode for the "Name" Data?

Organize

- Mode is 4.

Reflect

- The mode is generally not affected by extreme scores. The reordering may have a marginal effect.

Starting 1-7: Measuring the Spread of the "Name" Data

Organize

- Range = 5; IQR = 1.5.

Reflect

- The range would be greater. IQR could be affected by the reordering of the data.

- Range is affected by extreme scores. IQR is generally not affected. The reordering may have a marginal effect.

Starting 1-8: Who's Far Out?

Communicate

- Outliers must go *beyond* the fence—at the fence is not an outlier. Neither 1 nor 6 are outliers. Visually they both appear to be outliers.

Developing 1-1: Games Won

Organize

- The stem-and-leaf plot is as follows:

(Games won in 1993 by American League teams)

```
6 | 8  9
7 | 1  1  6
8 | 0  2  4  5  5  6  8
9 | 4  5
```

Legend: 6|8 = 68

Developing 1-2: Runs Scored

Communicate/Reflect

- It is possible to have two-digit stems or even two-digit leaves, but they may not produce effective displays. Rounding or truncating numbers reduces three-digit numbers to two-digit numbers.

Developing 1-3: Truncate or Round?

Organize

- The stem-and-leaf plot is as follows:

Truncated

```
6 | 7  8  8  9
7 | 1  3  3  7  8  9
8 | 2  3  4  9
9 |
```

Legend: 8 | 4 = 840–849

Rounded

```
6 | 8  8  9  9
7 | 2  3  3  8  9  9
8 | 2  4  5
9 | 0
```

Legend: 8 | 4 = 835–844

Reflect

- Truncated and rounded digits have the same level of precision.

Developing 1-4: Comparing Runs

- The legend is 73 | 8 = 738

American League				Stem	National League		
				58	1		
				59			
				60			
				61			
				62			
				63			
				64			
				65			
				66			
			5	67	2	5	9
		6	4	68			
			3	69			
				70	7		
			5	71	6		
				72	2		
		4	3	73	2	8	
				74			
				75	8	8	
			6	76	7		
			6	77			
				78			
			0	79			
				80	8		
				81			
			1	82			
			5	83			
			7	84			
				85			
				86			
				87	7		
				88			
			9	89			

Organize

Communicate
- Use terms such as *symmetrical, skewed, mound-shaped, rectangular,* and so on.

Developing 1-5: Boxing Games Won

Organize
- Lower extreme is 68.

  Lower quartile is 71.

  Median is 83.

  Upper quartile is 86.

  Upper extreme is 95.

Communicate
- About 25 percent won more than 86 games; about 75 percent won at least 71 games.

Developing 1-6: Comparing Games Won

Organize
- Lower extreme is 59.

  Lower quartile is 67.

  Median is 82.5.

  Upper quartile is 94.

  Upper extreme is 104.

Reflect
- A stem-and-leaf plot allows data for individual teams to be identified. This is not the case in a box plot.

- For comparisons involving more than two sets of data, the back-to-back stem-and-leaf plot would not work.

Developing 1-7: Runs Scored and Games Won

Reflect
- The data show a strong linear relationship that is positive in the sense that as the number of runs scored increases, the number of games won also increases.

Developing 1-8: Names and Shoe Lengths

Reflect
- Answers will vary, but generally the linear correlation will be close to 0, for example, there is no correlation.

Extending 1-2: Who Was the Top Female Tennis Player?

Communicate
- Students should include in their reports information regarding the lower extreme, lower quartile, median, upper extreme, upper quartile, and any outliers they determined.

  Lower extreme is $437,588 (Monica Seles).

  Lower quartile is $473,717.5.

  Median is $663,318.

  Upper quartile is $996,899.5.

Upper extreme is $2,821,337 (Steffi Graf).

Interquartile range is $523,182.

Outliers are $2,821,377 (Steffi Graf) and $1,938,239 (Arantxa Sanchez Vicario).

Reflect
- The omission of Seles data and the inclusion of Capriatti data would have no effect on the median, mode, and interquartile range. It would have a significant effect on the mean.

Extending 1-3: The Top Ten Roadsters

Communicate
- Students should include in their reports information regarding the lower and upper extremes, lower and upper quartiles, any outliers they determined, median, mean, and mode.

Reflect
- If only six were used, the range would be reduced and in this case central measures such as the mean, median, and mode would be increased.

Extending 1-4: The Speeds of Animals

Reflect
- A box plot would be the best way to display the data, however, a stem-and-leaf plot would also work nicely.

Lower extreme is 9.

Lower quartile (Q_1) is 29.

Median is 35.

Upper quartile (Q_3) is 44.

Upper extreme is 70.

Extending 1-5: Endangered Animals

- This information can best be contrasted with two box plots (using an asterisk (*) to indicate each outlier). Because there are large outliers, a stem-and-leaf plot would not be an efficient way to contrast the data.

Reflect
- The outlier would be at the lower end, instead of the upper end.

- It would not really affect the way the data were displayed, except that the box would move up due to the large numbers and the outlier would be at the lower end of the box plot.

Note: This activity illustrates that data are not always nice and neat nor do they always generate obvious and regular patterns. There will be outliers and unusual data. The whole aim of describing data is to show the general picture as well as the extreme cases.

Module 2: Probability, Simulations, and Random Numbers

Starting 2-1: Flip the Coin

Reflect

- Where a small number of trials is carried out, the experimental probability can be quite unstable. The effect of using the class data and the class mean provides a much more stable probability. This illustrates the Law of Large Numbers.

Starting 2-2: Model the Coin Flip

Reflect

- A coin is considered to be symmetrical and, for that reason, the theoretical probability of a head or tail is assumed to be $\frac{1}{2}$. In the experimental situation, one can get a considerable spread away from a 50 percent probability, especially when the number of trials is small.

Starting 2-3: Modeling Two Flips

Reflect

- Given that the number of trials is still relatively small, another 30 trials would normally give a different result. Heads could be assigned to the even numbers and tails to the odd numbers. In the long run, the probability of two heads should be 0.25.

Example

Outcome	**HH**	**HT**	**TH**	**TT**
Frequency	6	7	9	8
Experimental Probability	$\frac{3}{15}$	$\frac{7}{30}$	$\frac{3}{10}$	$\frac{4}{15}$

Starting 2-4: Steffi's Serve

Reflect

- Responses will vary. Make certain the trials are grouped correctly—a trial is one serve if the first is good and two serves if the first is a fault. The theoretical probability of a double fault is 9 percent, which is a good indicator of what students might find for the experimental probability.

Starting 2-5: Flip the Cup

Reflect

- Responses will vary depending on the shape of the cup.

Starting 2-6: Chances for the Name

Reflect

- The sum of the probabilities is 1.
- It would not affect the sum of the probabilities, however, the number of events would change.

Starting 2-7: Modeling the Chances for the Name

Reflect
- If Stephan-Lannell joined the class, there would be a very small sector to represent the probability of his name being drawn. It is very likely that a different class would have a different spinner, although there would tend to be a preponderance of events having a small number of first-name consonants.

Starting 2-8: Two Broadway Favorites

Reflect
- There are four outcomes. Probabilities are: seen *Phantom of the Opera* and seen *Les Miserables* (0.6); seen *Phantom of the Opera* but not seen *Les Miserables* (0.2); seen *Les Miserables* but not seen *Phantom of the Opera* (0.1); not seen *Phantom of the Opera* and not seen *Les Miserables* (0.1).

Starting 2-9: Three in a Row

Reflect
- As an indicator, the theoretical probability is $(0.6)^3 = 0.216$.

Developing 2-1: Three Hits

- In order to accommodate the three-decimal probability, you would need to use random three-digit numbers, for example, from 000–999. As an indicator the theoretical probability is $(.325)^3 \approx 0.034$.

Reflect
- The assignment of random numbers would be unchanged. In order to accommodate four consecutive hits, you would need a string of 12 random numbers.

Developing 2-2: The Top Ten

Communicate
- If p is the probability of the album being on top in a given week, the theoretical probability of it being on top at least two times in the next three weeks is: $3p^2(1 - p) + p^3$; at most two is $(1 - p)^3 + 3(1 - p)^2 + 3(1 - p)p^2$. In our case, *Music Box* has a probability of 0.6, hence $p = 0.6$. The answers are 0.648 and 0.784.

Reflect
- Although the question is different, the same simulation will answer it.

Developing 2-3: On Top for Four Weeks

Communicate
- As an indicator, the theoretical probability is p^4. This will be $(.6)^4$ if *Music Box* is six out of ten.

Reflect
- Although the question is different, the same simulation will answer it.

Developing 2-4: Basketball Playoffs

Communicate

- 0.55

Reflect

- It would not have affected the simulation, just the way you determined the experimental probability. A simulation is not needed; the probability of at most seven games is 1.

Developing 2-5: Grand-Slam Match

- 5

Communicate

- 0.72

Reflect

- It would not have affected the simulation. Different question, same simulation will work.

Developing 2-6: Chasing the Circle

Reflect

- Responses will vary.

Developing 2-7: Estimating π

Reflect

- The probability of hitting a circle equals the total area of the circles over the total area of the card. Since the dimensions of the card and the radius of the circle are known, the probability equation (probability = $\frac{\text{(total area of circles)}}{\text{(total area of the card)}}$) can be used to solve for $\pi \approx 3.14$.

Extending 2-1: Bazuka Bats

Communicate

- As indicators, the theoretical probabilities are:

  p (0 hits) = 0.21

  p (at least 2 hits) = 0.39

  p (4 hits) = 0.01

Reflect

- If his batting average changes, you would have to reassign the random numbers to fit the new probability. For example, if his batting average dropped to .320, the digits from 000 to 319 (or some equivalent) would be assigned to a hit.

Extending 2-2: Rock-Star Cards

- The expected number of boxes should be approximately 12.

Reflect

- For seven rock star cards the expected number is approximately 18. In the simulation you only need seven numbers to assign to each of the cards. Ignore three of the numbers—for example, 8, 9, 0, or some other equivalent assignment.

Extending 2-3: Raining in Surfers' Paradise

- If Sunday is a rainy day then the assignment of random numbers for Monday must be changed accordingly. Two spinners could be used, each divided into ten equal parts. The "follow a rainy day" spinner would have a probability of rain at 0.6 and the probability of a sunny day at 0.4. The "follow a sunny day" spinner would have a probability of rain at 0.8 and the probability of a sunny day at 0.2.

Module 3: Counting

Starting 3-1: Naming the Dog

- Students should notice that the problems in this activity involve mutually exclusive groups. Hence, addition is appropriate.
- 26

- 692

Starting 3-2: How Many Faces?

- There are 30 outcomes that have sums that are greater than or less than 7.
- There are 24 outcomes whose sums are even or less than 6.

- This problem involves the *fundamental counting principle*. If the first stage can be done in x ways, and the second in y ways, the number of ways of completing the two-stage task is xy. This can be generalized for more than two stages.

  (1) 88^2

  (2) All of them

  (3) $(1 * 88) + (87 * 1) = 175$

Starting 3-3: Birth Month in a Box

- In this problem, students should notice that to be certain of having two people with the same birthday month, you have to accept the possibility that all 12 months might come up before one month is repeated. This is known as the *pigeonhole principle*.
- 13

- 367 (leap-day birthdays are included)

Starting 3-4: Three the Same

Reflect
- The results are $12(n - 1) + 1$, where n represents the number of names with birth month alike.

Starting 3-5: First Names in a Box

- 14

Reflect
- 15 (assuming there were no other Mikes in the class)

Starting 3-6: How Many Sandwiches?

- This problem also involves the fundamental counting principle. Hence, multiplication is appropriate.
- 6

Reflect
- 6,480

Starting 3-7: How Many Kinds of Pizza?

- 24

Reflect
- 4,032

Developing 3-1: Batting Order

- $3! = 6$

Communicate
- $9!$

Developing 3-2: Pitcher Bats Last

- $8!$

Reflect
- $7!\ 2!$

Developing 3-3: Raffle Tickets

- 24

Reflect
- $n(n - 1)(n - 2)$; $10 * 9 * 8 = 720$.

Developing 3-4: Calculating the Number of Raffle Tickets

- $_6P_4 = 360$

Reflect
- Responses will vary. Make certain students' problems do incorporate *ordered* arrangements.

Developing 3-5: "Special" Raffle Tickets

- $9 * {_9P_4} = 27{,}216$

Reflect
- $_{10}P_5 - 27{,}216 = 30{,}240 - 27{,}216 = 3024$

Developing 3-6: "Extra-Special" Raffle Tickets

- $(9 * 8)(_8P_3) = 24{,}192$

Reflect
- The five-digit number is preceded by an ordered arrangement of the three letters A, B, and C with no repeats. Also, zero cannot appear as one of the first three digits.

Developing 3-7: Mini-Lotto

- 6

Developing 3-8: Calculating the Mini-Lotto Pairs

- 6

Reflect
- $_4C_2 = \frac{_4P_2}{2!}$

Developing 3-9: Pick Three

- $_4C_3 = 4$

Communicate
- $6; 6 = 3!$.

Reflect
- The key point is to have students recognize that $_4C_3$ is equal to $\frac{_4P_3}{3!}$

Developing 3-10: Pick Three Again

- $_4C_3$

Communicate
- $6 = 3!$

Reflect
- $_nC_r = \frac{_nP_r}{r!}$

Extending 3-1: Many the Same

- $366, 731, \ldots; 365(n-1) + 1$

Reflect
- Answer for 3 would be 730 if February 29 were the only restriction.
- at least 1461
- Missing days and student numbers could affect the solution.

Extending 3-2: Two in the Same Month

Communicate
- $12^5 - {_{12}P_5}$ (the total number minus the number where all five are different)

Reflect
- The probability that at least two of the students have the same birthday month is $\frac{(12^5 - {_{12}P_5})}{12^5} = 0.62$.

Extending 3-3: The State Lottery

Reflect

- The probability for one ticket is 1/25,827,165.

- Yes because you have two chances. Your probability is 1/12,913,583. In the Australian Lottery the probability for one ticket is 1/8,145,060.

Extending 3-4: Winning Hands

- $\frac{(_{13}C_1)(_4C_3)(_{12}C_1)(_4C_2)}{_{52}C_5} = 0.00144$ (full house)

Reflect

- 0.00024 (four of a kind); 0.002 (flush)

Module 4: Modeling and Predicting

Starting 4-1: Shoe and Arm Lengths

Communicate

- The points approximate a straight line with positive slope.

Reflect

- When students are predicting from the line of best fit, they need to understand that they can do this graphically or, at a later stage when they have an equation, analytically. They should also notice that they can predict y values for x values that are contained in the data as well as for x values that are not in the data, but are within an acceptable range.

- Responses will vary for the prediction of student arm length corresponding to a shoe length of 35 cm. Approximately 56 to 60 cm.

Starting 4-2: Shoe and Name Lengths

Communicate

- This data set is not linear. It is very difficult to fit a straight line. Many points would not fit well.

Reflect

- Shoe length and arm length have a strong linear relationship. This is not the case for a shoe and name length.

Starting 4-3: The Missing Movie

Communicate

- The scatter plot indicates a strong linear relationship. Note that as "rank" increases "money made" decreases.

Reflect

- Each student should draw the line of best fit and use the line to find the point were $x = 5$. This will give the value for *The Empire Strikes Back*. Responses will vary since each student will probably have created a different line; however, responses should be between 140 and 150 billion dollars.

Developing 4-1: Shoe Lengths and Consonants

Reflect

- The intention of the *Reflect* section is to focus students' attentions on errors that arise when a line of best fit is constructed. The errors will be least when data tend to be a good linear fit and will be greatest when they are not a good linear fit. The use of the absolute value in $|y - y'|$ ensures that the magnitude of the error can be used in calculations. This avoids having error values such as 5 and (–5) negating each other.

Developing 4-2: The Line of Best Fit

Communicate

- The mean square error will tell the student how close the points are to the line that they formed. The smaller the mean square error you find, the better the fit of the line.

Reflect

- The intention is to let students see that the mean square error varies depending on the location of their line. The "best fit" is the line with least mean square error. Later, students will see that the TI graphics calculators find the exact "least-squares" line.

Developing 4-3: Equation for the Line of Best Fit

Organize

- Equation of the line of best fit (regression line) is $y = -0.004x + 3.780$, values rounded to three places.

Communicate/Reflect

- The regression line produced by the calculator has the least mean square error. It is called the least-squares regression line.

Developing 4-4: Graphing on the Scatter Plot

Communicate

- Because the data are so scattered, there is not a strong linear relationship between x and y. Hence, any line that is drawn will not adequately represent the relationship between the x and y variables. In fact, the regression line is approximately horizontal, that is, its slope is 0. It predicts approximately the same y for every x.

Reflect

- There is no correlation, so the linear correlation index is close to 0.

Developing 4-5: Finding the Correlation

Communicate/Reflect

- Because $r = -0.01$ there is essentially no linear relationship between the two variables. The relationship is negative but this is of little consequence since r is so close to 0. The 0 correlation reflects the poor fit of points about the least-squares regression line.

Developing 4-6: Money Made and Movie Ranking

Communicate

- Correlation coefficient = −.936 using all the data except *The Empire Strikes Back*. The correlation is very strong because it is at −.936, which is close to −1. As the rankings get larger, the profits go down; the negative correlation coefficient reflects this.

Reflect

- The line of best fit will be very close to the points because the correlation coefficient is very close to −1, meaning there is a strong linear fit.

Developing 4-7: Money Made and Year Released

Communicate

- The correlation coefficient = −.741, again using all data except *The Empire Strikes Back*. The correlation coefficient is moderately strong because it is halfway between .5 and 1. The negative sign shows that there is a decrease in money made when the years increase.

Reflect

- The correlation coefficient for the movie ranking—money made is −.936. The correlation coefficient for year released—money made is −.741. Both correlations are negative, but the movie-ranking correlation coefficient is much stronger and hence more indicative of a linear relationship.

Developing 4-8: Forecasting the Missing Movie

Communicate

- The *y* value should be around 197 billion dollars. One interpretation is that the movies of the 1970s and 1980s have had more time to accumulate greater profits. It is also possible that movies of these decades were more popular.

Reflect

- The students should use the line-of-best-fit equation to find the movie's profit for the current year. When the movie was produced in 1994 (*x* = 1994), the top movie's profit should be around 128 billion dollars. The more recent movies make less money as was explained above. This will affect any predictions based on years beyond 1990.

Extending 4-1: Shoe Lengths and Arm Lengths

Communicate

- The predicted value of the arm length will be approximately 43.39 cm. The prediction is a good one because you are using a line of best fit with a strong correlation. However, there can be cases where extreme values are not predicted well.

Reflect

- As in any other data, predictions made from extreme values are often subject to the most error.

Extending 4-2: Testing the Radius and Area Relationship

Communicate

- Using a scale for x between 0 in. and 3 in. marked off in tenths of an inch, the scatter plot should look like a parabola. Another measure that will help you test for linearity is the correlation coefficient; it should be around .97 (if the radius of the circles are measured precisely, for example, to nearest tenth of an inch).

Reflect

- $A = \pi r^2$ or "the area is proportional to r^2"

Extending 4-3: Transforming the Radius and Area Data

Communicate

- Again, the appropriate measure is the correlation coefficient, which is about .99. The linear correlation coefficient of the transformed data (x, y_1) is stronger than the original data (x, y) because in the transformed data you are essentially making the area proportional to the square of the radius.

Reflect

- Yes.

Extending 4-4: Fitting Radius to Area

Communicate

- The linear regression model for (x, y_1) is $y_1 = 1.8x + 0$. Because $y_1 = \sqrt{y}$ this means that the (x,y) model is $\sqrt{y} = 1.8x \rightarrow y = (1.8)^2 x^2 \rightarrow y = 3.24x^2$. The answers should be around $y_1 = 2.7$ and $y = 7.29$.

Reflect

- Yes, it is the same relationship.

Extending 4-5: Testing the Radius and Volume Relationship

Communicate

- The process of transformation is similar to that used in Extending 4-4. Students should take the cube root of y instead of the square root in order to make the volume proportional to the radius.

Reflect

- The two values should be approximately the same providing the data have been accurately collected.

Extending 4-6: Crude-Oil Production

Organize/Communicate

- The scatter plot should look like a slanted S. The correlation coefficient is .8746. This coefficient is strong and positive; however, the scatter plot suggests that the data are not linear.

Reflect

- The scatter plot indicates that a nonlinear function would provide a better fit than a linear function. The mean square error for the linear model is 16,180,737.

Extending 4-7: Transforming the Crude-Oil Production Data

Communicate
- The scatter plot is essentially unchanged by this transformation. This is also true for the correlation coefficient, which is 0.8711. Residual plots and the mean square error (16,598,827) confirm that the transformed model is actually worse than the linear model.

Reflect
- No; see the discussion above in Communicate.

Extending 4-8: Fitting Year and Crude-Oil Production

Organize
- $y = 402074.34\, x_1 - 3036415.34$
- $y = 402074.34\, \ln x - 3036415.34$

Communicate
- $x = 2000 \rightarrow y = 19712$

Reflect
- Both processes should lead to the same answer since $x_1 = \ln x$.

Extending 4-9: An Alternative Transformation

Organize
- $y_1 = 0.062x - 111.911$
- $\ln y = 0.062x - 111.911 \rightarrow y = 2.5 * 10^{-49} * e^{0.062x} = 2.5 * 10^{-49} * 1.06^x$

  The transformed scatter plot shows a good linear approximation.

Communicate
- $x = 2000 \rightarrow y = 75587$

Reflect
- The correlation coefficient is .990. The residuals are more random but the mean square error is 50,597,769. Exponential transformation is much better but the end points blow up the mean square error.

Extending 4-10: Another Transformation

Organize
- $y_1 = 119.263x_1 - 895.317$
- $\ln y = 119.263\, \ln x - 895.317$

Communicate
- $x = 2000 \rightarrow \ln y \rightarrow 11.191 \rightarrow y = 72443$

Reflect
- The correlation coefficient $= .9911$ indicating a strong linear relationship for the transformed data. The mean square error is 45,570,736 and the residual pattern is less random than for the exponential transformation in Extending 4-9. It is a much better fit than the transformation in 4-8 and marginally worse than the transformation in activity 4-9.

Extending 4-11: Non–English-Speaking Americans

The table below presents comparative information on linear and nonlinear regression models.

Regression Model	Equation	Scatter Plot	Correlation Coefficient	Residuals Plot	Mean Square Error
Linear	$y = 0.909x + 102553.17$	Appears somewhat linear. Line is reasonable fit.	0.938	Residuals somewhat random	$3.496 * 10^{10}$
Natural Log $(x_1 = \ln x)$	$y = 352634.74\ln x - 3828675.11$	Transformed line is not a better fit.	0.844	Systematic pattern	$8.34 * 10^{10}$
Exponential $(y_1 = \ln y)$	$y = 141315.34 * 1^x$	Transformed line is not a better fit.	0.822	Fairly systematic pattern	$10.507 * 10^{10}$
Power $(x_1 = \ln x)$ $(y_1 = \ln y)$	$y = 13.761 * x^{0.809}$	Transformed line is as good as linear model if not better.	0.936	Residuals somewhat random	$3.401 * 10^{10}$

The linear model and the power model would provide the best predictions.

LINEAR: $x = 11{,}549.333 \rightarrow y = 10{,}604{,}046$

POWER: $x = 11{,}549{,}333 \rightarrow y = 7{,}163{,}475$

Extending 4-12: How the United States Has Grown

The table below presents comparative information on linear and nonlinear models. In order to reduce the size of numbers, the *x* variable has been transformed so that 1790 becomes 10, 1800 becomes 20, and so on. 1990 is 210 under this transformation. This is a linear transformation so it does not affect the shape of the scatter plot.

Regression Model	Equation	Scatter Plot	Correlation Coefficient	Residual Plot	Mean Square Error
Linear	$y = 1.214x - 47.825$	Appears exponential in shape. Line is reasonable fit.	0.960	Systematic curved pattern	458.3
Natural Log $(x_1 = \ln x)$	$y = 77.180 \ln x - 258.744$	Transformed line is worse fit. Transformed data no more linear.	0.804	Systematic curved pattern similar to linear residual plot	2073.0
Exponential $(y_1 = \ln y)$	$y = 4.740 * 1.021^x$	Transformed data appear more linear. Line seems to be better fit.	0.983	Different curved shape from linear and natural log, but still systematic	1194.7
Power $(x_1 = \ln x)$ $(y_1 = \ln y)$	$y = 0.045 * x^{1.557}$	Transformed data somewhat more linear. Line seems to be reasonable fit.	0.971	Curved residual plot similar to linear plot	560.4

In this case the natural log transformation is not suitable, based on several criteria. The linear and the other two non-linear models are reasonable but certainly not satisfactory on all criteria as the table shows.

When $x = 2000$

LINEAR: $y = 219$ million

EXPONENTIAL: $y = 458$ million

POWER: $y = 198$ million

These predictions suggest that the linear model is best, at least for making future predictions.

References

1. Department of Mathematics and Computer Science, North Carolina School of Science and Mathematics. *Data Analysis.* Reston, Va.: National Council of Teachers of Mathematics, 1988.

2. Dossey, John, et al. *Discrete Mathematics.* Glenview, Ill.: Scott, Foresman and Company, 1987.

3. Kenney, Margaret J., ed. *Discrete Mathematics Across the Curriculum, K–12:* 1991 Yearbook. Reston, Va.: National Council of Teachers of Mathematics, 1991.

4. Landwehr, James M., and Ann E. Watkins. *Exploring Data: Revised Edition.* Palo Alto, Calif.: Dale Seymour Publications, 1995.

5. Moore, David S., and George P. McCabe. *Introduction to the Practice of Statistics.* New York: W. H. Freeman and Company, 1993.

6. *Organizing Data and Dealing with Uncertainty,* rev. ed. Reston, Va.: National Council of Teachers of Mathematics, 1979.

7. *Curriculum and Evaluation Standards for School Mathematics: Connecting Mathematics,* Addenda Series Grades 9–12. Reston, Va.: National Council of Teachers of Mathematics, 1991.

8. *Curriculum and Evaluation Standards for School Mathematics: Data Analysis and Statistics,* Addenda Series Grades 9–12. Reston, Va.: National Council of Teachers of Mathematics, 1992.

9. Newman, Claire M., Thomas E. Obremski, and Richard L. Scheaffer. *Exploring Probability.* Palo Alto, Calif.: Dale Seymour Publications, 1987.

10. Gnanadesikan, Mrudulla, Richard L. Scheaffer, and Jim Swift. *The Art and Techniques of Simulation.* Palo Alto, Calif.: Dale Seymour Publications, 1987.

11. Travers, Kenneth J., et al. *Using Statistics.* Menlo Park, Calif.: Addison-Wesley Publishing Company, 1985.

12. Witmer, Jeffrey A. *Data Analysis: An Introduction.* Englewood Cliffs, NJ: Prentice-Hall, 1992.

Other books of interest:

Barbella, Peter, James Kepner, and Richard L. Scheaffer. *Exploring Measurements.* Palo Alto, Calif.: Dale Seymour Publications, 1994.

Burrill, Gail, ed. *From Home Runs to Housing Costs: Data Resource for Teaching Statistics.* Palo Alto, Calif.: Dale Seymour Publications, 1994.

Burrill, Gail, ed. *Teaching Statistics: Guidelines for Elementary Through High School.* Palo Alto, Calif.: Dale Seymour Publications, 1994.